COMMENTS ABOUT *BUILD IT TWICE:*

"This is truly a valuable resource for anyone involved with construction projects. On an even broader scale, since 'building' could be a metaphor for a journey through any landscape peopled with both friendlies and hostiles, it can be a resource for planning any kind of project, business and even personal. The reader will find not only the substance of *Build It Twice* valuable, but that its style is engaging, humorous, logical and consistent. I'm looking forward to reading it again and again as I'm presented with the various traps and pitfalls that are almost inevitable in construction."

-Steven Baltzell, V. P., DPIC

(Design Professionals' Insurance Corporation)

"*Build It Twice* exposes and explains the rough realities of the construction process. The reader is reminded that you can cheat an honest man and that good intentions are not a substitute for personal responsibility. But the author fails as a cynic—because he offers the reader a detailed, clear and practical roadmap, guide, and arsenal of tools with which to drill through the stressed-concrete jungle. There is hope. Anyone who is contemplating building anything larger than a bird feeder *should read Boyd's book twice.*"

-James G. Luce, Esq.

"The genius of this book is that, without minimizing the realities of the conditions that can be encountered in the development of a construction project, it provides strategies that can benefit everyone associated with that project. *Build It Twice* demonstrates how both the novice and the experienced owner of a building project can define and implement a plan that, even in an industry where secret alliances have been in the making for decades, will deliver a win/win situation for everybody."

-Malcolm A Misuraca, attorney

"This is a unique contribution to the literature of project development. I am unaware of any other effort that has taken as comprehensive and realistic a look at the hazards of the industry, and none that has followed up that look with a realistic strategy the owner can employ to avoid most of those hazards.

"It is a book that anyone concerned with delivery of a successful project—owner, architect, contractor, attorney—would do well to read more than once."

-David Baldwin, director of litigation support systems, Ibis Consulting

"Having begun my legal career representing owners on everything from strip shopping centers to nuclear power plants, I am familiar with the tendency of owners to rely exclusively on their designers until it was too late. Joe has made this fact abundantly clear in his book.

"He has also opened windows for the owner into some hidden areas of construction disputes (contrary to Bismarck's advice that there are two things men should know nothing about: the making of laws and of sausage).

"Owners—and others in the construction industry—*should* know about the things Joe Boyd describes in *Build It Twice.*"

-Gregory Thomas, construction attorney

Joe Boyd's

Build It Twice

...If You Want
A Successful Building Project

Split Rock Publishing
Napa, California 94581

Copyright © 1999 by J.A. Boyd

Published by Split Rock Publishing Company
Napa, California
Printed in the United States of America

Book Design: Beth Rodda
Publishing Consultant: Linda F. Radke, Five Star Publications, Inc.
Editor: Sal Caputo, Five Star Publications, Inc.
Cover Design: Lynlie J. Hermann, Five Star Publications, Inc.
Proofreader: Gloria Casas
Illustration: Michele Le Blanc

Library of Congress Cataloging-in-Publication Data
Boyd, Joe
        Build It Twice ...If You Want a Successful Building Project / by Joe Boyd.
        1st ed.
        p.   cm.
        ISBN 0-9663620-0-4
        1. Building—Planning. 2. Construction Industry—Planning.
        3. Construction contracts. I. Title
        TH438.B665 1999
        690 ' .068—dc21   98-35010
                                                                        CIP

Requests for permissions should be addressed to:

Split Rock Publishing Company
P. O. Box 5517
Napa, CA 94581
e-mail: bit@splitrockpub.com

For Michael, who is still here, contributing and debating.

INTRODUCTION

WHERE ARE WE GOING?

1

PROJECT SUCCESS

2

CONSTRUCTION

3

FOREIGN TERRITORIES

4

DESIGN

5

THE LAW

6

TIME: THE IRREPLACEABLE RESOURCE

7

HOLLOW PROMISES AND MORE ABSURDITIES

8

NEW TRACKS

9

YOUR PROJECT PLAN

WHERE HAVE WE BEEN?

APPENDICES

INTRODUCTION

Every calling has its secrets. Since money changes hands based on these secrets, every calling also has its ambushes and its little ways of defeating legitimate expectations. This is true even if a contract, or even the law, say otherwise. It is often true even when those who don't know the secrets, or don't know whether they are being manipulated or abused, retain experts to protect them.

The construction business is among the worst offenders and the chief exponents of the iron rule that *what you don't know will probably hurt you*. The reason is rooted in the fact that when a new owner—someone who has never built anything before—first arrives on the scene, the players already assembled on the field have alliances and customs in place that will take advantage of the owner's inexperience. This can be true even of those who are nominally the owner's agents, those who are supposed to protect him from his inexperience.

Who's included within this quasi-indictment? All the usual suspects— the architect, the contractor, the subs, the inspector, and even (or maybe, especially) the construction manager hired to keep the architect and the contractor on the straight and narrow.

It's pointless in an introduction to try to rationalize this perennial circumstance, and certainly there is nothing in this book that remotely tries to justify it. Joe Boyd's deep commitment to an ethical and principled construction game is clear from the first lines of the book. When he speaks of "project success," he means the win/win where everybody receives what they are entitled to receive, and no one makes an extra dollar at the expense of the legitimate rights or expectations of anyone else, even a rookie owner.

The point of this introduction is to consider how much a neophyte has to know going into this book for the first time. Can this or any other book protect a rookie from the lessons that have been accumulating, the alliances that have been in the making, for many decades? Does an owner need to be overwhelmed by what it takes, by what he must know, to survive in such an environment?

The answer is the genius of this book. The book could only have been written by someone like Joe Boyd, who, despite 40 years experience with these realities, has not lost his boyhood sense of right and wrong. I happen to know somewhat whereof I speak, because Joe and I are both products of a Jesuit education. Once through that, you never quite get out of your system the sense that there's a right and wrong to be observed in every encounter in life.

But it would be a grave mistake to consider this book either a long complaint about how things ought to be, or Pollyanna's answer to these realities. This book takes the construction business as it is, as Joe Boyd has found it over four decades. What it does not do is let the game be any less fair than it can be when someone on behalf of the owner (like the guide Joe suggests) has an eye constantly on what's going on, often in ways invisible to the untrained eye.

If you are a new owner, therefore, new that is to building something for which an architect, a contractor, an inspector, a construction manager and others will converge on your property and begin doing things, this is the book for you.

It won't be easy, this book. It will challenge you every chapter, often to read and then re-read a chapter, and sometimes to put the book down and return to it later, maybe many times.

Is it worth it? Is this a book that an inexperienced owner can hope to master? The simple answer is: only if you want to know the truth and to protect yourself from it. If you can't afford the effort this book requires, you can't afford to build what you intend to build without paying some unexpected, and possibly unjust, dues.

If you're an owner involved in continuing building programs, you can hardly afford not to study it.

Can you master it? Probably not in the sense of getting everything that is here in the first pass, or even the second. But then, you'll have weeks or months probably, while you're planning, designing, bidding, and then building. Whatever time you have before you, spending that time picking up this book a third or fourth time will be an investment that pays dividends each time you revisit it. It is a book that will yield something different each time, that reveals itself in a different way depending on the reader's immediate need for information or the issue that is at hand. (Joe has provided a study program for this in chapter 23).

For each part of the book Joe Boyd has given you a paragraph or two to orient the chapters that follow, to suggest what's in store and at stake in the following chapters. Venture forth only when you've read that heading. A cookbook or a check list it is not. If you lose track of the direction you're headed, read the orientation again, or go to the chapter entitled WHERE HAVE WE BEEN? Stay with it.

For myself, I expect to own this book as long as I am involved in construction. As a lawyer, I have a special need for it, to keep me from persuading myself that the law always knows how to deal with construction cases. That's not true, and I plan to spend many a long hour sifting through this book to keep me on the straight and narrow.

Malcolm A. Misuraca

WHERE ARE WE GOING?

We will be visiting the unfamiliar territories of the strangers an owner must meet and work with in the development of building projects. In the course of these visits, we'll discuss an idea that has consistently eliminated most of the problems those strangers can create for an owner's project budget.

WHERE ARE WE GOING?

You'll find here an idea that has delivered successful building projects for over 40 years.

Every type of project: libraries, hospitals, homes, manufacturing plants, auditoriums, submarine bases, jails, aircraft maintenance bases, high rises, shopping centers.

For all kinds of owners: corporations, medical facilities, cities, universities, school districts, counties, individuals, public agencies and private organizations.

The idea has been employed productively by business since the Industrial Revolution.

It's been at the core of defensive-driving courses for years.

Software development firms have made it policy for program development.

And yet it's a rare idea in building construction.

Most owners have never been introduced to the idea; and wouldn't know where to start if they had been. Their design consultants aren't motivated to tell them.

Few owners have had the chance to consider the idea because a zone of silence has evolved in the gap between the owner and the cottage industry that exploits opportunities on projects to which it hasn't been applied.

And because it requires a specific kind of fortitude, commitment and perseverance on the part of the owner to span that gap and to carry out the programs the idea requires.

It's an idea that, effectively implemented, can eliminate 80% of the construction problems that decimate project budgets and invite construction claims. Problems that result from defective construction documents.

It's an idea that, effectively implemented, can eliminate 80% of the construction problems that decimate project budgets and invite construction claims. Problems that result from defective construction documents.

3

It's an idea that, by enhancing the owner's awareness of the interrelationships between design, construction and the law, can help an owner reduce the other 20% of construction problems, too—the 20% that can't be entirely foreseen.

It's not a new idea. In fact, it's a throwback to the way buildings were built hundreds of years ago.

It's an idea that exploits the latest management technologies.

But it has nothing to do with the buzzwords, cant phrases and panaceas that are offered to deliver projects *on time and on budget while avoiding claims.*

It takes more work and more commitment than those hollow promises deliver.

It also takes a willingness to make some changes in the way things are typically done in design and construction.

The idea is simply to analyze and construct the project on paper while it's being planned and designed—to analyze the project progressively during design at a level of detail equivalent to what contractors will undertake while bidding and building it.

To implement this idea, it's necessary to visit the territories of those contractors in order to see this future project as they will see it.

Then, it's necessary to double back and visit with the design consultants, to see what's required to facilitate this progressive program of *building it before it's built.*

And it's important to visit that peculiar territory known as *the law,* to learn how it views the whole matter of projects that get into trouble, and what can be done to keep them out of it.

We'll be taking a journey through these territories—the territories of contracts administration, programming, design, bidding, construction, and construction-claims resolution.

4

When you return from our trip through unfamiliar territories, equipped with the resources for your real trip, then you will have a decision to make.

- Either to follow the conventional tracks that have produced a 1,000 percent increase in construction claims in the past 25 years.

Or

- To lay out your own tracks, and to change the way your own project will be measured, managed and documented.

Come along and see what it takes to build it twice, and the values it can bring to your project. Then, you can make your decision.

Stranger: Someone you would not allow to control your checking account.

The Territories, the Strangers and the Vocabularies of Construction

On this trip we'll visit several of the territories from which strangers will arrive to design and build your project. Each territory has its unique vocabulary and unique practices. We'll hear some strange tongues. We'll also hear some cant phrases and buzzwords that have no real substance.

We'll describe these phrases as we come upon them, but if you want a preview you can visit the glossary in the appendices.

We'll also visit some of the territories from which *other strangers* might arrive if your project finds itself in the arenas of *alternative disputes resolution.* More strangers than you ever want to meet.

How To Get There?

I've started our journey by introducing you to some thought processes general contractors pursue behind closed doors, and especially out on the job—because that's where most owners first take notice of the consequences of theirs and others' inattention to details.

Then we'll go to where too many trips end—in a courtroom. This very brief visit is just to suggest how a simple idea can be misunderstood, misrepresented and clouded in legal mystique.

But you may want to start your trip in Part 4, or Part 5 or Part 7. Each part of this book can be approached as a stand-alone introduction to a particular territory.

At some point, though, you will begin to consider which option you intend to pick up of the two I suggested on page 5. As you consider these options you may want to go to the end of the book, to WHERE HAVE WE BEEN? That's where you'll find the practical index to this book.

There is no conventional index at the back, because when you double back—as I expect many of you will do frequently—you'll be looking for ideas, not just words.

That's how BEEN is organized—around ideas with which you can define the separate components of your Plan as it becomes a cohesive direction for your project, your building Program and your organization.

Construction Gender

Man was only a short distance from *his* cave when *he* built *his* first shelter. As a result the construction industry is bound to a lot of male-generated words. *Manlift, manpower* and *workmanship* are permanently with us. Their politically correct substitutes are awkward and useless in the context of getting a better handle on how to deliver a successful project. Rather than compromise the effort to convey ideas, I've made only modest efforts to adapt them.

To maintain some balance in considering the owner's profile, though, I've sometimes used *"his or hers."* But since *his/hers/its/ theirs* does nothing for clarity, I'll ask your indulgence by reading "he," "she" or any other gender interchangeably into the usage that best fits your need or disposition.

Joe Boyd,
The Napa Valley, 1998

1

PROJECT SUCCESS

Success for a building project is in direct proportion to the respect that flows back and forth between the owner and those who design and construct it, respect that results from the integrity of the construction documents and the consistency with which the Contracts for those services are managed.

Establishing and maintaining a consistent standard throughout the development of a construction project requires clarity of ideas, fortitude and commitment to a well-defined direction—the same characteristics that in every undertaking separate success and failure.

1

WHAT IS SUCCESS FOR A BUILDING PROJECT?

Project success is the subject of this book.

In the broadest sense, success for a building project means:

- That the resources available for the project's development deliver their optimum value.

- That the project stays out of trouble, which today pretty much means that it avoids the demolition derby known as a "construction claim."

Through the rest of this book we'll be looking at ways to assure project success from an owner's point of view.

Before we do, it's useful to spend a moment considering the broader view of *project success* described above. It points up a significant fact about construction.

> *When an owner delivers construction documents that make it possible for everyone working on the project to maximize the use of the available resources, the chances are vastly improved that the project will be successful for everyone associated with it.*

When construction documents are carefully developed to protect a project's budget, those documents also will reduce the potential for disputes among the "strangers" the owner hires to build it.

As disputes are reduced, so is the potential for those "other strangers"—the ones who arrive with a construction claim—to drain the project's resources.

And surprise! When the owner ensures that those documents are managed effectively, he improves the odds that a core of well-intentioned people out on the job will help deliver a successful project to him.

There are competent people on every construction project—most of whom the owner will never meet—who have more leverage to keep project resources in line than the owner can ever hope to have. What these competent people respond to is a set of clear, well-coordinated, well-managed construction documents.

1

Every owner, along with his or her own definition of *project success,* needs to carry this broader view of success into the struggles he should expect while he's trying to ensure that those standards are incorporated into the development of his construction documents.

For an owner a successful project is one that's:

- Delivered within its original budget, expending no more than a reasonable contingency amount;

- Delivered within a schedule projected with sufficient detail to make it realistic; and

- Consistent with the owner's requirements for quality, aesthetics and function.

The unfortunate fact is that many projects don't meet any of the standards of success above.

Why?

That's the extended subject of this book.

The answers to that question will evolve as we visit those essential strangers you will be hiring on your project. We'll find why so many projects fail. We'll also discover how you can improve the odds for delivering your project within all of those standards for success.

The Buildability/Budget Analysis (B/BA)

The engine that drives successful projects is *anticipatory conceptualization:*

> *Visualizing future physical relationships and potential conditions far enough in advance to eliminate or reduce the problems they could generate.*

In order to develop the ability to visualize future conditions—to see things coming—you first need to know what all those strangers out there will be doing to bring those conditions about.

You need to know what they need to be doing to make the project successful—and the forces that are set in motion if they don't do them, or don't do them well.

A Short List of Successful Projects

Here's a short list of building projects that benefited from the effective implementation of the tools you'll be evaluating for your Project Plan:

- A building program required the construction of 40 separate projects, each on a different site, over a five-year period. Construction costs were reduced 18 percent and durations were reduced by 44 percent by the time the 10th project was completed.

- A single-family residence was completed well enough below the owner's original budget to pay for most of the landscaping and to add a swimming pool.

- Over three years, the engineering and construction cost overruns that had been experienced on a $850 million treatment plant expansion program were reduced by two-thirds.

- A $30 million, 400,000 S.F. industrial building, designed and built under a cost-plus-fixed-fee contract was completed on schedule with only 0.6 of one percent in design-caused change orders.

- Modernization of a 90-year-old hospital constructed originally of unreinforced masonry walls was completed on time and within budget, without disruption of critical hospital operations despite a complicated, phased program for seismic reinforcement.

- A $23 million justice center and detention facility was designed and built under four separate and sequential fast-track construction contracts, without disputes, on schedule, using only one-quarter of the contingency money provided for possible changes.

None of these projects experienced the agonies of a construction claim.

The process of conceptualization for an owner needs to start *before* the owner starts those strangers out of the gate. So he or she can define their scopes of work and the means to measure their performances.

The owner needs to fill in what I've called *the gap*—the owner's ignorance about what those strangers do.

As we explore *the gap* and start to describe what those strangers you hire should be doing on your behalf, you may start to lose sight temporarily of the title of this book: *Build It Twice.*

Where it all will come together is in the *Buildability/Budget Analysis (B/BA).*

1

I've coined the phrase *Buildability/Budget Analysis* to distinguish the process suggested in this book from a similar phrase used in the industry: *Constructability Analysis*. The reason for a new phrase is that *Constructability Analysis*, like so many other catch-phrases and buzzwords in the industry *("critical-path-analysis," "construction management," "partnering,"* etc.) have acquired bad names because they've been so underperformed that they often don't deliver what they promise.

So, just for the duration of this book, you have a new phrase that's intended to cover the owner's comprehensive conceptualization of *everything* that's likely to happen on his or her project. As a buzzword *B/BA* is nothing special. It's just a convenient interim name. Once you pick up on what it includes, it will not matter what you call it; only that it's effectively implemented on your project.

In one line:

> A *Buildability/Budget Analysis* is an effort by the owner to conceptualize <u>before</u> and <u>during</u> its planning and design all of the relationships—human, physical, fiscal, environmental and political—that will influence the development of the owner's project.

This is a sizable challenge, but it's no more than should be done on all projects. However, it usually isn't. Because most design consultants, despite their representations and their promises, can't produce such an analysis. And because the standard contracts they offer the owner don't provide for it.

A B/BA does call for the owner to acquire a different mindset from what most owners have, but the evidence we'll review in this book will demonstrate how important this mind shift can be.

We can't get into the processes of a Buildability/Budget Analysis, however, until we take a journey through what those strangers do, and what happens when they don't do those things effectively.

Then we'll roll up our collective sleeves and come back to discuss what an effective B/BA requires.

The analysis of hundreds of construction disputes has demonstrated that the single most significant element for the success of a building project is the quality of the construction documents.

Buildability/ Budget Analysis

A Buildability/ Budget Analysis is an effort by the owner to conceptualize <u>before</u> and <u>during</u> its planning and design all of the relationships— human, physical, fiscal, environmental and political—that will influence the development of the owner's project.

Many of the ideas in this book come from having assisted with the resolution of construction disputes over the past 30 - 35 years. The causes of those disputes are the anvils against which the ideas in this book have been tested and refined.

Those tests have demonstrated two important facts.

- First: reduce the causes for the worst kind of construction dispute—the *delay/disruption claim*—and the owner will eliminate up to 80% of all of the reasons projects exceed their budgets.

- Second: by reducing the causes of a delay/disruption dispute, an owner will almost automatically eliminate most of his concerns about quality, aesthetics and functionality.

A Different Mindset

Despite the evidence that the success of building projects is directly related to the quality of their construction documents, owners pay for all kinds of redundant services to keep contractors "honest" and to "avoid claims"—to keep contractors from exploiting the conflicts, ambiguities, omissions and outright errors the owner already has allowed to be published with his construction documents.

It's during construction—with the horse already out of the barn—that owners try to implement checks and balances. The owner hires "construction managers," invokes "partnering"; and retains claims consultants and attorneys to protect against liabilities that almost certainly will result once defective documents are issued.

There are tons of evidence (literally, in the weight of construction claims paper alone) that owners should be most concerned during design about what goes, or doesn't go into their construction documents. It's then that owners should be most skeptical; yet it's then that they are most trusting.

Owners typically accept out-of-hand what their design consultants tell them about the quality, the design coordination, the budget costs and the projected construction durations represented by the design documents.

Even though the proprietary agreements that consultants, contractors and construction managers offer typically contain totally inadequate check-and-balance systems, and the specifications

Delay/Disruption Claim

A delay claim asserts that the construction contract duration has been extended as a result of conditions beyond the contractor's control. A disruption claim asserts that the contractor has been unable, due to conditions beyond his control, to perform contract work as efficiently as he originally planned and bid it. These claims are often associated, but require distinctly different proofs to support their assertions. (See chapter 15.)

attached to them are 50 or more years out of sync with reality, owners readily accept them with only minor adjustments just to get things going.

It's before authorizing *any* of these agreements that an owner will most benefit from an observation Dr. Sam Johnson made 250 years ago:

> **"We are inclined to believe those whom we do not know because they have never deceived us."**

Since Sam's comment 250 years ago there's been considerably more evidence of our almost universal gullibility, but owners of building projects continue to hire design consultants and contractors based solely on their own representations, on shallow interviews, or on the recommendations of others.

These are the ironies we'll be visiting in the following chapters.

Whatever your program, private or public, if your direct concern is conservation of the money needed to build or remodel your facilities, then how to achieve *project success* is certainly one of your essential concerns.

The 3rd & 4th Dimensions

Some projects are so complicated that it's important to model them in three dimensions while they're being designed, to work out how irregular surfaces meet, to check and confirm volumes and interfaces of unusual shapes.

- Designers of refineries and processing plants frequently model in three dimensions the interfaces of the equipment needed to support complex operations. Large rooms are filled with multi-colored plastic models of piping, pumps, tanks, cranes, towers and conveyors.

- My firm, J.A.Boyd & Associates (JBA) has worked with contractors on projects to develop models of the temporary shoring and support structures for underground foundations, so we could work out the logistics of getting materials into and out of deep excavations.

- We have also built wood, plastic or cardboard models of projects to visualize and balance their shapes and volumes, and to adjust their orientation to winds, the sun and the environment. On several projects we had to build three-dimensional models because our clients—many of them public agencies with ongoing building programs—could not visualize building spaces from their two-dimensional drawings.

- With computer software it's now possible to fly around or walk through colored images and rooms of buildings before they're fully designed.

What the B/BA adds to these three dimensions is the 4th dimension of *time*.

Testing the buildability and construction logistics of a project's design against time can substantially improve its construction documents.

Allocating costs to the work activities that are defined in a detailed time analysis will progressively confirm its budget as design progresses.

The atrributes of a B/BA are described in detail in chapter 14, and in Appendix A-2.

Some Contemporary Reasons That
Project Success Is Important

Before we leave the matter of *success,* it's useful to consider why it's becoming more essential every day to concentrate on where your project funding goes.

Recent projections (1997) indicate that within the next decade the requirement for school classrooms in the primary and secondary grades will increase by 10 percent nationwide, which translates to increases of 35% across a band of southern and southwestern states.

If you have any contact with your local school board you already know how limited the funds are to bring school technology into the 21st century. What you may not know is how much of this money is being wasted through the archaic—and heavily lobbied—requirements of state educational departments. Much of the work to be done on schools relates to underground utilities and rough-in of services (gas, electrical, etc.). Architects are paid 9% – 12% for doing little more than stamping drawings, while the complete design *and* construction of these systems can be performed by engineers and engineering contractors for fees on the order of 5% - 7%.

If you're building a home or some other kind of project for yourself, your search for funds and project financing has told you more than any book could about the importance of watching where the dollars go.

If you're in private business, you know better than public agencies why it's important to conserve capital.

If you're on a public board and have seen projections for future projects, you are all too aware of the erosion of your tax base. You surely have a concern about how much of the money available is wasted. Lacking funding, some public agencies have entered into developer-leaseback agreements to meet the requirements of their charters for highways and detention facilities, but these programs aren't available for every public need. So you know the need to conserve and to be efficient.

As a hospital administrator you may, once you've taken this trip with us, begin to wonder about the wisdom of turning the direction of the hospital's development program over to a design professional whose capabilities in the areas of management, economics, construction and cost control may be suspect. There's hardly a hospital in the U.S. that wouldn't benefit from a more efficient way to design and construct renovations and additions.

2

THE RESPECT OF STRANGERS

Everyone who develops a building project wants it to be successful.

The reason so many building projects are *not* successful is that their owners carry their images of success to the point of visualizing their completed projects and stop.

Many owners get as far as how their project will look, how it will function, how much it will cost or how many widgets it will produce. Hospital administrators concentrate on services to patients. School districts are concerned about the curriculum and safety of their kids. Public agencies assess how many people a project will serve. Others plan how many offices, elevators, trucks, sales counters or apartments it will contain.

Then these owners turn their projects over to strangers.

Build It Twice (BIT) *is based upon the premise that:*

- *An owner who retains strangers to assist with the development of a building project must provide a clear, unambiguous plan for achieving the goals of that project.*

- *The owner, in order to make this plan work, must first acquire a comprehensive understanding of what those strangers will be doing as they interrelate to deliver that project.*

- *The owner, by applying this understanding to build the project on paper as it is being designed, can reduce 80% of the reasons projects are not successful.*

When you have someplace to go and you know you will need the help of strangers to get there, you need to gain their respect before you can ask for their help.

Once you gain their respect, you tap an unfathomable well of pride and support from the competent stranger.

The most effective way to gain that respect and support is to demonstrate that you have clear images of where you plan to go and how you plan to get there. That you recognize the

2

difficulties you and those you bring along will encounter. And that you're prepared to provide a clean, clear and accurate map of the journey beforehand.

When competent strangers receive that kind of map they usually carry the less competent along with them.

Build It Twice takes the owner's perspectives beyond *what* and *where* the project will be, to *how to get it there successfully.*

Build It Twice is based on over 40 years of evidence that supports this premise.

Over the first few years we were in business, our consulting firm (J.A. Boyd & Associates) assisted owners such as United Air Lines, Wells Fargo Bank and the Veterans Administration, plus several hospitals and universities, to deliver almost a hundred successful projects (there were 20 projects in UAL's program, 40 in WF's).

These projects were completed within their budgets (some below budget) and without exposures to construction claims. In 1971, we published a description of our management program. It was called *Build It Before It Starts.*

By then, we were also responding to the requests of owners, contractors and design professionals to assist them with the resolution of construction disputes on other projects.

The ideas in *Build It Twice* come from these combined experiences: helping owners and contractors build successful projects; and helping others dig out from the holes they'd gotten into on less-successful ones.

To build it twice, an owner needs to conceptualize the entire journey through which a project travels, from planning through design to construction.

The map of that journey, for a building project, consists of several parts: the agreements for design and construction services; the specifications that define scopes of work and performances; the drawings that represent your completed project; and the procedures that will hold the whole thing together until the journey's completed.

18

That's why *Build It Twice* goes beyond even those contracts that involve a contractor early on.

Build It Twice recommends that an owner who is committed to the success of his or her project trace the processes of design and construction back from the images of the completed project.

- Back from its completion all the way to the content of the agreements that define the performances of those strangers.
- Back to a point *before* any of those strangers are retained.
- Back to the development of a Project Plan that anticipates how those strangers will relate to each other and to the owner.

Build It Twice suggests that the owner, having accumulated some familiarity with the processes of design and construction, incorporate into his Plan a system that defines how he intends to measure, document and direct the performances of the strangers who will help him build it.

There are types of construction contracts that approach the ideas recommended in *Build It Twice*. Contracts that bring the experience and perspectives of a contractor into the design process (see chapter 19).

When these contracts have delivered successful projects, it has been because the owner put into place a system of checks-and-balances to measure the performances of the contractor as well as of the design consultants.

What distinguishes the ideas in *Build It Twice* from these contracts is the recommendation that the owner familiarize himself with the contract options available, and define—independently of the formulas those "strangers" offer—procedures that will inform the owner, over the life of the project, of the performances of all of the strangers he hires to design and build it.

The ideas and the tools you'll find described here don't need concepts like "construction management" or "partnering" to be effective. They won't require that you resolve a particular type of design or construction contract—lump sum, cost-plus-fixed fee, design/build or any of their many variations. The ideas here don't interfere with, or eliminate, any of these concepts or contractual formats either.

They're applicable to any kind of building project. They can be applied as effectively by a public owner as by a private one.

The purpose of this book is to provide the owner of a potential project with ideas that have worked on other successful projects, public and private; and to encourage that owner—you—to develop a Plan unique to your own project and organization.

Hello Broadway!

A Broadway play is a fair analogy for the ideas in *Build It Twice*.

> First, you write a script. You visualize the stage settings. You write stage directions. The only people in the room are you, your choreographer (your guide) and your musical director (your attorney).
>
> Once the play's direction is set, the actors arrive for auditions. You choose those who best fit the roles you've written. You provide the script to them (not the other way around).
>
> You go through weeks of rehearsals, getting the best out of the actors you've selected, benefiting from their talents—giving them their head when it contributes to the play, controlling them when it doesn't (because you know in advance where the play has to go).
>
> By opening night, you've lived through the play, tweaked it, cut down some scenes, expanded others, and you *know* it's a success even before the curtain rises.

It works on Broadway, and it's delivered the successful projects described in this book.

It works because it shares, along with the benefits of dress rehearsals, the tenets of an effective defensive-driving course:

Know your vehicle;

Know your capabilities;

Anticipate what others may do;

Set up your own responses to be in control of possible situations.

The Territories

The chapters that follow are based on the premise that the more you know about the businesses of the firms you retain to work with you and for you, the better job you'll get.

Your recognition of the obstacles and problems they face as they pursue their businesses and attempt to help you with yours should help you bury your predispositions and prejudices behind your larger motivation to produce a successful project.

Strangers

Strangers will arrive on your project from many different territories. Some of these strangers are essential to the success of your project. Others, you'd just as soon not meet.

Each *essential* stranger will contribute some piece of his or her territory—like the piece to a jigsaw puzzle—to the owner's project.

When—and if—those pieces are put together in the right order, they will become a landscape painting with the owner's project as its centerpiece.

In order for that to happen efficiently—to optimize the resources available—it's the owner's responsibility (literally and legally) to provide a clear map for those strangers to follow.

That map is the set of directions the owner needs in order to measure and control the development of the construction documents. In this book, I've called it the owner's Project Plan.

Mapping the Territories

In order to prepare that map, the owner needs to know where each of those essential strangers is coming from. He needs to know about the territories in which they work, what they do there, and how they relate to strangers from other territories. The owner needs to develop some familiarity with the languages they speak.

In the next few chapters, we'll visit the territories of these "essential strangers."

We'll also briefly visit some of the "other strangers," the ones the owner would just as soon not meet. We'll find them in the land-

2

scapes of projects that were less than successful—projects that were dragged all the way to the trial of a construction claim. These visits will be useful to demonstrate what can happen when the pieces of the jigsaw puzzle don't come together efficiently.

In the courtroom and in the respective territories of the consultants and contractors the owner hires, we'll come across foreign languages, buzzwords and even some common words that are used in unfamiliar ways.

These visits will demonstrate that commitments based on buzzwords and foggy phrases are among the worst sins of the construction industry. That these buzzwords often mean entirely different things to promisor and promisee. You may even come to conclude that "construction management" and "partnering" are among the worst offenders.

We'll see these loosey-goosey buzzwords and concepts in more than one context. We'll do what we can to demolish the illusions they try to create. When you're back home again, you should be in a position to do the same when you're presented with the hollow promises behind these acronyms and supposed panaceas.

The glossary in the appendices has some cross-references to the text that will bring you back to the broader discussion of a particular term. There are also some appendices that expand on the management tools we'll discuss, so you'll have a better idea of how they work before you pick them up to use them.

One of the purposes of this book is to help you learn something of the vocabularies of design, construction, project documentation and construction disputes; and of the legal processes that often frustrate the reasonable resolution of those disputes.

By the time we're back from our trip, you should be familiar enough with these vocabularies to challenge the concepts, the panaceas and the legal smoke and mirrors that can leave you in doubt about what's being done in the name of your project.

Delegation versus Abrogation

Anybody you think of as successful has spent a lot of time studying the territory they planned to go into before laying down their chips, for example: Nordstrom's Department Stores.

Few owners think about doing on their own projects the kind of homework that Nordstrom's, for example, does in researching its markets and selecting its inventories.

Even many organizations with continuing building programs know less about the contracts they sign and the duties they "delegate" than Nordie's does about shoes.

Although their project's success will be tied directly to how well the parties to those contracts perform their duties, what they do and how they do it is a mystery to most owners. This ignorance generates a predictable set of conditions:

- If an owner doesn't know what goes on in the performance of the duties for which he contracts, he can't measure them.

- If he can't measure the performance of those duties, he can't influence them.

- If he can't influence them, then those duties haven't been *delegated,* their administration has been *abrogated.*

The owner who executes a design, construction or administrative contract, while knowing less about what those parties do than Nordie's does about its shoe inventory, is likely to find all sorts of unwelcome strangers in his project landscape even before the paint dries.

What those strangers do, or fail to do, when they're passing through the owner's project landscape can have both immediate and long-term implications for that project's budget.

What they do, or fail to do, can obliterate the boundaries the owner has put around the first costs of construction, and they can leave behind conditions that can abuse a project's operating and maintenance costs. They can cause the owner nightmares for years after they're gone.

To avoid those nightmares, the owner of a potential building project needs to challenge some conventions.

One of those conventions is the acceptance of the standard agreement forms that architects and others offer. Our visits will demonstrate the hazards hidden in these agreements.

The owner who plans to challenge the status quo, however, needs to define a replacement. To do that, he has to understand what those consultants and contractors should be doing on his project to make it successful. He has to understand their duties. And he needs to refine their scopes of work, because the standard contracts that are out there don't do an adequate job of it.

He needs to know how those 30, 50, 100 or more firms will interact to produce his project.

He needs to know how to measure and document what they do in order to *be ahead* of the problems he wants to avoid.

Murphy's Law and the Other 20%

Hard-rock granite would have to be blasted out to build an access road to the new project that would be built on the other side of the military post. A hillock of trees about which the military was justifiably proud stood to one side of the planned roadway. The contractor was specifically directed that the trees were to remain undisturbed.

When dynamite is used to blast rock, the blast-master spaces his charges so that the charges go off microseconds apart, starting at the point farthest away. This sequence breaks up the rock progressively, so the force of successive blasts is relieved from the area to remain intact.

On the first day of the construction, the entire post, military brass, band and all were assembled to celebrate the occasion.

While the blast echoed across the area and 400 pairs of eyes watched, the trees—and the hill on which they had stood— all disappeared in a cloud of dust. The blast-master had reversed the blast sequence.

3

SPANNING THE GAP

Controlling Project Resources

The U.S. in the last several years appears to have gotten control of rampant inflation. We have gotten a lid on the conditions that led so many organizations, and even countries, into catastrophic real estate losses just a few years ago.

Business and industry have responded to the need for managing resources more efficiently with waves of improvements in management techniques. Industry, transportation, technology, retail and service areas have all benefited. Across-the-board improvements in productivity have reinforced this control.

The benefits of this balanced economy have stabilized the costs of construction materials and equipment, and labor costs also appear to have leveled out.

In the area of construction, though, there are forces hidden to most building-project owners that are operating with different motivations.

The forces that lurk beneath the surface of many projects can be insidious. They create potentials for strangers to exploit any opening the owner provides with the distribution of a project's construction documents. These forces allow strangers to exploit the very resource the owner is trying to conserve—the project budget.

These forces are *the other strangers in the gap.*

The Gap

The gap: the space between an owner's images of his or her building project and a clear picture of what it takes to deliver it. Or the space where the owner should have developed a heightened awareness of the many ways a building project can get into trouble.

The gap has been evolving over the last 150 years as technology has improved and enhanced the design and construction of building projects. We could not build the projects we build today 150, 50, or even 20 years ago. But, as with most technology, these advances have brought trade-offs with them.

3

As industrial and technological advances have improved construction, they've also inserted more strangers between the owner and his or her project. The number of strangers required to design and construct a project has increased substantially over the past 30 years and exponentially in the last 15 years.

The central problem—the one we're discussing in this book—is that neither the contracts nor the management procedures for administering projects have kept pace with these technical advances. Neither have they been refined to respond to the environmental, political and cultural demands and changes that can influence project development. These agreements do not address the need to manage all the human strangers those advances and changes demand.

Instead, the documents that define project administration have evolved in another direction. They have added more strangers—human and ideological—to the mix, neither enhancing the owner's awareness of project liabilities nor protecting the owner from them.

The associations—the AIA, the AGC, the NSPE, etc.—that provide the "standard agreements" for their members to offer owners, have made these contracts more and more protective of their members', not an owner's, concerns for project liabilities.

AIA: The American Institute of Architects

AGC: The Associated General Contractors of America

NSPE: The National Society of Professional Engineers

As these associations have been confronted with the demands for inclusion of additional participants in these agreements—"construction managers" for example—they have added more standard forms to accommodate them. These expanded forms often confuse, duplicate and dilute the services for which the owner believes he's paying.

The proprietary agreements these associations present to an owner invite redundancy and duplicative costs, have dissipated authority and responsibility, and have provided more escape hatches for their members than a nuclear submarine.

The means to measure or document failures of performance have been so diluted in these agreements and specifications that when things hit the fan, everybody brings a legal action against everybody else. It's inevitable that these legal actions involve the owner.

All of this is encouraged by the legal profession, scrupulously guarded by the sureties, and exploited by the "disputes resolution" cottage industry that has attached itself as a parasite to construction.

Protecting the owner's budget is not among the primary motivations of these "other strangers."

The Ideological Strangers: Buzzwords and Panaceas

As technology has widened the gap between the owner and his perceptions of *what goes on out there*, communication shortfalls among the participants on the owner's project have increased. So have failures in contract performances. This has had the logical effect of generating ever-larger spaces between the owner and the strangers he hires to build his projects.

As Parkinson predicted, these voids have increasingly been filled with more strangers. They include not only the human strangers looking for their own places in the gap, but the management themes—the ideological strangers—they promote.

"The team concept," "Construction Management," "Partnering," "TQM," "Segregated Prime Contracting," "Value Engineering" are just a few of the buzzwords, panaceas and relationships promoted to deliver projects "on time and on budget" while "avoiding claims." Panaceas to address the problems the last strangers contributed.

These buzzwords are often only hollow promises. Ironically, the acronyms that are used to promote them ("TQM" for example) have been taken directly from business and industry, where the concepts they represent have been very successful.

They've been gathered up by the construction industry without qualification and presented to unsuspecting owners as panaceas for construction problems, problems they typically do not solve because of the inherent nature of construction itself. Owners have accepted them because they sound like solutions that have been successful in other businesses.

But they seldom work on a building project because of the inherent nature of construction.

TQM - Total Quality Management

A concept conceived in the U.S., rejected by U.S. Industry, enthusiastically adopted by Japanese industry as a means to incorporate top-to-bottom control of quality and productivity into its manufacturing and marketing efforts. This approach to management of industrial production has been re-imported into the U.S., and is now espoused by the construction industry. (See chapter 16.)

27

3

Construction is a one-time assemblage of strangers, each driven by motives which, if they aren't quickly directed toward the owner's goals, will run out of control, which is the first sign of another construction dispute about to happen.

Business and industry are essentially horizontal activities. To increase production in a factory you can install more lathes and drill presses. You can add scientists and programmers to increase the productivity of an R&D, production or programming project. These measures work because these projects are essentially open-ended. They can be expanded horizontally.

Building construction is a vertical, transient activity with a limited life. It is limited spatially (you can't build the 3rd floor before the 2nd). It is limited by the transient relationships of the resources that can be invested in a construction project: time, manpower, equipment, money and competent management.

Construction is a one-time assemblage of strangers, each driven by motives which, if they aren't quickly directed toward the owner's goals, will run out of control, which is the first sign of another construction dispute about to happen.

Meanwhile, a "disputes resolution" cottage industry is standing in the wings, waiting to rush into the gaps where these cures have failed, bringing with them their panaceas and more strangers than any owner ever expected to meet.

R&D - Research and Development

Research and development falls somewhere between vertical and horizontal projects, in that resources can be increased, and scopes of work subdivided, within the limits that the direction and the results of research allow.

This cottage industry is still evolving to take advantage of the 1,000% increase in construction disputes and claims that the U.S. has experienced between 1965 and 1995[1].

Asking Questions

Considering how many budget-destructive forces are waiting out there in the gap, it's no revelation that many owners don't know what kind of questions to ask, or when and where, to ask them. The owner isn't sure who to ask, or whether he can trust the answers he gets.

The owner often doesn't know what those strangers out there are supposed to be doing to produce a successful project. The information he receives is usually insufficient to measure whether they're doing those things. He seldom knows whether those strangers are performing up to the promises in their contracts. He usually has only one way to respond—he stonewalls to protect his budget.

[1] *Along with providing project-consulting services over the past 40 years, the author has also been part of this cottage industry, so the insights expressed here come from both sides of the matter.*

Even when things begin to hit the fan, it's usually the owner who's the last to know what happened, or what the longer-range implications could be.

What makes the difference, ultimately, between a successful and a less-than-successful project is the owner's ability to *anticipate:*

- To ask meaningful questions before the conditions that present those questions become project liabilities;

- To recognize responsive answers and to be able to integrate those answers into an effective action; and

- To take effective actions in a timely manner.

There is no better way to recognize hollow promises than to demand timely, responsive answers to challenging questions from the people making those promises.

It is the owner's ability to anticipate that makes the difference between a successful and a less-than-successful project.

A Plan, a Notebook and a Guide

You'll be accumulating many of the questions you need to ask those strangers as we get deeper into the territories from which each one will arrive. You'll also have your own questions about where one stranger interacts, or one idea fits, or doesn't fit, with another. My goal is to have answered most of those questions by the time we've completed our trip.

Putting those questions down in writing will provide a long-term benefit. It will start you toward what is another essential goal of your journey: developing a Project Plan.

Why a Guide?

Because of a variation on the Peter Principle.

You may or may not have been promoted beyond your competence to grow into the management position you occupy, but it's almost a certainty that you will be pushed beyond your experience when it comes to the variables that can influence a building project.

And, unlike any other position you may ever hold, a construction project is not a time or place to learn on the job.

3

It's inevitable, however, that you will have more questions than I can answer in a book. So I suggest that right about here is a good time for you to start a project notebook—one in which you can record the questions you'll want to ask those strangers, your own questions about how things work, and the answers and ideas that come to you as we cover more of what is a convoluted, complex topography.

The notebook will enhance your facility to ask directed questions and to demand responsive answers from those you hire to help you build your project.

The notebook will help you develop the Project Plan that is the final goal of your trip. Your Project Plan will be unique to your project. It will be driven by your unique goals, limited by your resources and encouraged, or restricted, by the organization, firm or agency with which you work.

Because your Plan will be unique, and because your current perspectives of design and construction activities are probably limited (or you might not be reading this book), you may need a personal guide to help you interpret and apply these ideas to your special circumstances. That's another good reason for your own personal project notebook. It will be a useful workbook you can share with your guide once you find him or her.

We'll discuss the profile of an ideal guide in Part 9. One essential quality your guide should have is that he or she should be able to answer most of those questions you've collected—and to help you ask the challenging questions you will want to ask those strangers you will be hiring.

2

CONSTRUCTION

You and your contractor will initially see one another as strangers. His first impression of you will come from your construction documents. On the basis of those documents you will exchange promises.

Whether, and how, those promises are fulfilled will determine how many "other strangers" you will meet.

4

THE CONTRACTORS

First Impressions

The first time a contractor hears about your project, he's off in some other owner's landscape, or several landscapes at the same time.

He's involved with the logistics of other projects, dealing with other owners, architects, building departments, inspectors, his subcontractors and suppliers and possibly with attorneys, claims consultants and sureties.

Excepting for the owner and contractor who have a continuing relationship, the first, critical impression the contractor will get of your project will come from your construction documents (CDs).

The contractor will know in one day if those documents are clear, consistent, coordinated and buildable—or not.

He's probably already familiar with the work of your architect. He'll know some, or all, of the other consultants. And if you have one, he'll know how your construction manager operates. He'll probably know all of them better than you do.

The Contractor and the Bid Period

Once a contractor becomes aware of the opportunity to bid your project, his first decision is to determine whether it fits into his current business plan, his available bonding capacity, his resources and workload and whether he's prepared to invest the several thousand dollars it will cost him to prepare an estimate and bid.

You've provided four weeks for return of bids. Half the time will go by before he invests much effort in studying your bid documents. He'll probably discuss the project documents with favored subcontractors who may start dissecting them before he does. They'll come by his plan room and sign in, so he'll know who's reviewing them. He can piggyback on their impressions.

Even if he is interested, the general contractor (GC) will probably defer getting into your CDs until the last week or 10 days before the bid is due. But he'll be getting impressions from various sub-

4

contractors, handling other projects, considering whether yours fits his workload and whether his staff has time to estimate it. Or whether some other project fits better into his plans.

The contractor's attention and response might be modified if he's one of a handful on your selected contractor list, or if he's been contacted beforehand by you or your architect. But his close study of the documents will be concentrated into the last week before bid date, while his potential subs feed him their own perspectives.

When he does start to bid your CDs, his first effort is to assign his staff to prepare "quantity take-offs," which are tabulations of the work *he* expects to perform, as distinct from what subcontractors will do. He will review the specifications with one eye to the support equipment and construction logistics he will be required to provide, and those that subcontractors will provide.

"CPM scheduler"

Usually an idependent consultant whose unique talent is the ability to draw "critical-path network diagrams" and process them on a computer.

His or her competence in construction logistics is frequently notable for its absence.

He will parcel out assignments for the insurance, bonding and scheduling requirements. As likely as not, he'll solicit bids from a "construction-CPM scheduler" (an oxymoron if ever there was one), because the largest percentage of building contractors do not have this peculiar kind of stenographer in-house. He will be interested in getting the lowest bid for this, but whether or not he in fact pursues the matter of scheduling even if he is awarded the contract will still be a matter of speculation and of the owner's commitment to the contract specifications.

The contractor's estimating staff will assign unit costs to the bid quantities he has already accumulated, using his in-house historical costs for similar work, an estimating manual, or by checking with his viscera.

On the bid date, there will be a fire drill in the contractor's office, while his secretaries, estimators, junior engineers and his brother-in-law take phone bids from subcontractors (right up to the last 11 minutes). He will then phone the bottom line into someone waiting with a cellular phone near the owner's office. This person will plug the number into the bottom line; lick the envelope and drop it into the owner's depository with 45 seconds to spare[1].

In that last hour of receiving telephoned subcontractor bids, the general contractor's management team has selected the lowest

---

[1] *This is typical of many public bids under lump-sum contracts. Procedures become less hectic and judgments more refined as the owner and contractor have more opportunity to study each other's psyches and vibrations.*

sub-bids, plugged in WAG (wild-ass-guess) numbers where bids were not received, and made a final *really* SWAG pass (*scientific* wild-ass-guess) by adjusting the bottom line to meet his visceral sense of the competition.

> *Even if the process of preparing and assembling a construction bid is somewhat more civilized, it's absurd to believe, as some interpreters of the AIA A201 construction agreement suggest, that the successful bidder has carefully catalogued the conflicts and defects in the construction documents and incorporated the costs of their corrections into his bid.*
>
> *Or that he has worked out the logistics of the project in much detail.*
>
> *Unless you believe that aardvarks can fly.*

AIA—A201

The standard AIA contractor form for a construction contract resolved on the basis of a lump-sum (single amount) cost quotation for the Work. The 1987 version, although it has been subsequently modified by the AIA, is still in wide use.

Addenda and Changes

Changes to the construction documents, after they're printed, but before bids are submitted, are produced in the form of addenda. An addendum is a supplemental document, produced by the architect, that changes what's in the bid documents, the specifications or the drawings. Addenda are issued during the bid period and become part of the original construction contract. Changes made as addenda must be incorporated into the contractor's bid. Changes after bids are submitted can only be made as change orders (contract modifications).

With the issuance of the first addendum, the first red flag is raised to the contractors:

> *"The owner's already changing the documents. What'll it cost us to adjust our bid, confirm with subcontractors? How many more afterthoughts will there be if we get the job? Let's take a closer look at how the documents are coordinated. Looks like there could be more changes coming."*

The Search for Change Orders

Some contractors may assign an engineer or estimator early in the bid period to research the documents for potential change orders. Contractors who typically bid public work, who are already familiar with the performances of the offices that produce these documents, are particularly tuned to this potential.

4

RFI-Request for Information

Sometimes, RFC (Clarification). A request that flows from the contractor to the design consultants, raising questions about information in the construction documents. The process started by an RFI can end with a competent, complete answer; or can go all the way to a disputed contract modification.

An owner who does not make an investment in careful analysis of the construction documents, their budget and their contract duration before they go out for bids is a sitting duck.

The general contractor is a businessperson, in business to make a buck. If he can, he'll do that by developing a bid that's responsive to the bid documents, a bid that allows reasonable and adequate costs for doing the work.

If he can't—if the documents are ambiguous, incomplete or unbuildable—he'll use any other means that will assure him he's not going to pay for the owner's failure to issue clean drawings.

Or to compensate for ambiguous specifications.

The lack of an adequate management plan will be telegraphed to bidders in many different areas of the specifications:

- The lack of controls over the timeliness required of the architect and his consultants' in response to submittals and RFIs;

- The documentation, or lack of it, required for change orders and time-extension requests;

- The format of the payment reports;

- The scheduling and reporting requirements;

- The meeting-reports requirements;

- Inadequate definition of excusable delays (weather, for instance);

- The documentation associated with scheduling and daily performance by construction forces;

- The documentation required from the architect, the contractor and/or the CM in relation to the processing of submittals, RFIs, change orders;

- Etc., etc.

There are no less than 15, and can be as many as 30 - 40 articles of the administrative specifications in which the mechanics for measuring, reporting and documenting performances will telegraph to contractors how loosely, or closely, the owner intends to oversee, measure and document his project.

The owner can also expect that the contractor will take into consideration the character and past performance of the owner, the past products of the architect, the competence and objectivity of the construction manager. To make these judgments, he has available not only his own experience, but the experiences of all the others he knows in the industry.

The Construction Industry: A Tight Community

Most owners forget, or don't know, how close a fraternity the construction industry is. The Associated General Contractors of America (AGC) is a national organization, with effective branches in every state. Many general contractors are members, and they also belong to a lot of other associations organized around their specialties and their communities: roadbuilding, homebuilding, dam building, mechanical and electrical trades, cost estimating, etc., etc. In addition to these, there are hundreds of local contractor and subcontractor organizations. The Internet has made the interchange of information and the checking of past projects and the latest regulations even more accessible.

There's a camaraderie in the industry that's not matched by design consultants. There is a sharing of information and perspectives that owners and professional associations can't even approach. A lot of golf is played, a lot of fish are caught every year by friendly contractor-competitors.

If your architect, your mechanical engineer, your construction manager or any other representative you retain has a reputation as an idiot, it'll precede him even as the contractors are bidding your construction documents.

The contractor has been down this road before. If he's been in the business at least 5 years, he's tried at least half the tricks in the book, and knows about the others from his friends in the industry.

Does that mean that all contractors are bloodthirsty and ravenous or lack integrity? No.

Excepting for the scurrilous and most opportunistic, most contractors would rather work with a clean set of construction documents intelligently and fairly administered than spend their time estimating and arguing change orders. Reputable contractors do not plan to make money on changes. If they had a reasonable quote going in, they probably had a profit figured they'd be willing to take away from your project. They'd prefer not to have to call their attorney.

CM-Construction Manager

Originally intended to be an independent and objective representative of the owner's concerns for administration of the construction contract, the concept is now so loosely used and widely exploited as to mean as much, or as little, as the ignorance of the owner and the opportunity to make an extra fee provide. (See chapter 16.)

4

The essential thing for the owner is to ensure that the contractors get those clean, coordinated documents, together with an intelligent, fair and objective program to manage them.

A program that keeps the design consultants as honest and as responsive as the owner expects the contractor to be.

Most general contractors want to work with the construction documents with minimal disruption to their overall business plans, make the profit they expected to make, keep their subcontractors working cooperatively toward the same goal, and move on to the next productive project.

But every contractor has known the bad project, knows where to look for problems and has the resources to mount a tough campaign if it means protecting his profit and his bonding capacity. His bonding capacity and his financial resources are his assurance he will be able to work, and keep his people working, tomorrow and next year.

The owner who allows a bad set of construction documents or ineffective administration to interfere with the contractor's business is inviting trouble from a contractor (and his attorney), who is better equipped to handle that trouble than most owners, architects or construction managers.

The First Challenge

Unless, with the construction documents themselves, and with an effectively structured bid period, you have telegraphed that the owner is fully in charge of where the project is going, you can expect that the contractor's been thinking:

> "We'll find an opening to test the owner's awareness and commitment. Are the architect's answers clear and timely, or evasive? Is the CM's staff green behind the ears? Do these guys know which side is up? Do they appear to be reasonable, or arbitrary? How soon should we put them in their place?"

You can bet that the contractor is thinking these thoughts during the bid period and in the pre-bid meeting. And again in the pre-construction meeting. You would, too, if you were about to put $2.5 million of your bonding capacity on the line.

Whatever answers the contractor's antenna receives, he will keep to himself how he plans to approach *this* project until the agreement is signed.

It should be no surprise to an owner that the first candid and challenging remarks from a contractor who has just been awarded a contract may not be raised until after the contract is signed. The contractor probably won't bring it up until after the first schedule and payment request are submitted, or the need arises for the first change proposal.

Then the train leaves the station. Momentum and gravity take over. Construction becomes a force that can't be stopped, or even slowed, without extending both the tracks, and the owner's exposure to extra costs.

The March of Time

The first moves are the contractor's. He transmits:

- Shop drawings and field drawings for early work: concrete, utilities, foundations.
- Submittals for long lead items: structural steel, hardware, or elevators.
- The project schedule and his version of the payment breakdown.
- Preliminary Lien Notices (unless the project is on publicly owned land).
- His first requests for information (RFIs).

"What does this mean? This doesn't match that. It isn't buildable this way... please tell us what to do."

Submittals, Shop Drawings

The construction documents define *what* the contractor is to build. Many of the details about *how* are left to his selection and discretion (bolting of joints, selection of equipment, etc.). The CDs require the contractor to submit these details in the form of drawings, or specifications and samples, which the design consultants must review and *accept* before they can be installed. (*Approve* used to be the operative word but has passed into oblivion, along with other concepts that raised the anxieties of designers and their insurers.)

The contractor is still saying "please." The marriage is still in the honeymoon phase. The first probing to test the responsiveness of the architect and the owner's determination is just starting.

The End of the Honeymoon

The first sign the honeymoon is starting to unravel can be one of many things. Some examples:

- The contractor's RFIs (requests for information) disappear into the architect's offices for too long.

- Or they come back with unclear or inadequate answers.

- Or they come back with an Architect's Supplemental Instruction (ASI) to change a design defect in the documents *with no additional time or compensation to the contractor.*

- Or the contractor presents a stack of EWOs (Extra Work Orders) at any early progress meeting.

- Or a contractor submittal is inadequate, and a dispute arises.

- Or an architect, insecure in his design, rejects the contractors' submittals for arbitrary reasons.

EWO—Extra Work Order

Contractors will often start a list of work they believe may lead to extra work that will prove to be compensable under their contracts. The contractor may or may not set up separate cost codes to track associated costs. EWO lists may be substantive, or specious, depending upon the contractor's attitudes about reasonable compensation, or about creating a potential claim environment.

ASI—Architect's Supplemental Instruction

This is a standard document, with standard small print on the back, issued by the architect for the purpose of directing the contractor how to proceed with certain tasks under the construction contract. The small print states that it is not a change order and does not entitle the contractor to additional compensation or extension of time to his contract. If the architect tries to disguise a change in scope or a contract modification in an ASI, the owner can expect a squawk from the contractor.

- Or the architect or CM accepts an inadequate schedule and payment breakdown and the project moves ahead to the first change order without bases for measuring time or dollars.

The list could go on for pages. Any one of many possible events can be the first sign of trouble.

The issues at the heart of all of them are whether the contract documents were clear about how time will be measured, negotiated, controlled, documented; whether the CDs contain the leverage the owner needs to measure and enforce them; and whether responsive contract administration will be maintained to disentangle the *"He said, you said, but I said"* arguments that will follow.

In the context of the messages the contractor's antenna receives, the central issue is whether the architect's and consultants' responses are timely, direct, responsive, complete, clear, objective and knowledgeable.

Interactivity

The interactivity owners are typically concerned with during construction include:

- Payments to the contractor; and
- Control of costs when changes are required.

But these are *reactions,* not the interactions that count.

The interactivities the owner needs to be apprised of—needs to be pro-active about—are the ones that occur *before* payments are due, before there are changes that impact the project:

- The interaction between the contractor, the construction documents, the design consultants and the construction manager (CM) (if any) during the bid period.
- The interactions between the contractor, architect and CM when those documents contain problems.
- The interactions between the contractor, his subcontractors and the architect when submittals, RFIs or issues of delay are processed.
- The interaction between the architect and its subconsultants that are required to respond to all of the above.

Impact

In construction, the word impact *is used differently from the way it's used in casual conversation. In construction lingo, it can be either a noun or a verb. It is used to identify a condition that is asserted to have interfered with the performance of the contractor. See expanded definition in the glossary.*

We will cover these and other interactions in more detail in the following chapters.

The Owner, One-on-One
With the Contractor on Small Projects

Sometimes, the owner will contract directly with a contractor to undertake all the work, including design coordination. The kitchen remodel, the small office renovation or a home addition are typical examples.

The owner hires the contractor directly and works out design requirements with him without benefit of independent consulting or other assistance.

It can be an area fraught with hazards. The same potentials we've already discussed are all there. They're not less likely just because it's a small project. The cost exposures may be smaller in absolute amounts, but can often be proportionately higher in relation to the size of the project than on larger projects. It's not unusual for these types of projects to cost 100% more than the owner planned—or worse, be left in a state that costs several times the original budget to remedy.

If you think about it, all the same rules apply as on other projects. They just have to be applied in a different way.

Your Program is as critical on a small project as on a large one. As much time and effort as is practicable should be invested by the owner into building scrapbooks, talking with other owners who have gone this way before, talking particularly with others who have worked with this contractor, and having someone trustworthy check his contract, his cost proposal and his plans.

It is important to collect as many sources as possible with which to define the terms and conditions you will need in your contract. Books on building and contracting are good sources to start with, as are the standard industry contracts (such as the AIA A201 for general contracting). Keep in mind that you'll still need to make many of the modifications we will be discussing.

Finding someone (your guide) to help you work out all the details before committing to anyone can be the most important thing you will do, even on a small project. You're likely to need an honest, objective, experienced source of advice along the way to complete your small project successfully, and your guide can prove to be a valuable resource.

Postscript

We could delve into the broad profiles and types of general contractors who might build your project. Contractors with "A" or "B" licenses. Contractors whose only office is their pickup truck. Contractors who have never operated a computerized cost report, prepared an adequate schedule for their subs, or contractors who keep no records of actual costs. (See *The Glove Compartment Contractor,* page 248.)

Contractors you may have had recommended by a member of your church or country club. Contractors you find in the Yellow Pages. Contractors who are as honest as you are, and contractors who don't know what a scruple is.

We could discuss specialty subcontractors, such as roofers, electricians, or dirt movers.

We could talk about the qualifications you should look for in a potential contractor or subcontractor, and the agencies and business bureaus you can check with before hiring, or even considering, a contractor.

There are plenty of areas we haven't covered in this chapter, some of which will come up in later chapters. But there are still likely to be areas in which you'll want more information before you can put all the pieces and parts together into a Project Plan.

If you have never contracted for construction or remodeling services before, the most important decision you can make will be to find that special resource I've called the owner's *guide.* His or her profile and capabilities are described in chapter 22.

Your guide should be able to help you anticipate the promises, scopes of work and performance you should require from any contractor on your small project.

5

THE OWNER AND
THE CONTRACTORS

Complete, Accurate and Buildable?

When did an architect last advise an owner that the construction documents the architect produces will deliver *a warranty from the owner to the contractor* that those documents will be *complete, accurate and buildable?*

The die is cast once the drawings go out to the blueprinter.

Unless the owner has assured their buildability and accuracy, it will be only a matter of time before his budget is threatened.

Or his project experiences a construction dispute.

It was the owner who authorized the issuance of the bidding and construction documents. It's the owner's landscape—the landscape in which 40, 60, 100 or more strangers will converge to deliver his project.

If the owner delivers anything less than a well-coordinated, effectively administered set of construction documents to those strangers, they will be preparing a surprise party for the owner before the ink is dry on the construction contract.

Despite the fact he's the one with the most to lose, it's the owner who knows least about what's in those documents. It's the owner who least anticipates how they can expose him to delays, cost overruns or claims. As a result, the owner's invitation to a surprise party is repeated on project after project:

> *"What are we doing here?"*

Why is the owner surprised?

Because there's a singular paradox in the construction documents. Neither the people who produce them nor the attorneys who review them take the trouble to tell the owner about it.

When did an architect last advise an owner that the construction documents the architect produces will deliver a warranty from the owner to the contractor that those documents will be complete, accurate and buildable?

5

The Owner's Warranties to the Contractors

The paradox is that the documents with which the owner guarantees the project to be buildable, are the same documents that contain most of the seeds of confrontation with the contractor.

It's also a paradox—if not something worse—that the consultants who plan, program, design and budget what's in those documents don't advise the owner about the guarantees he's making.

In case the reasons behind this paradox aren't yet obvious, the following is a partial listing of the conditions the owner warrants as his documents go out to the blueprinter.

- The owner has a duty to disclose any superior knowledge he has that the contractor will need in order to perform under the terms of the construction documents. If the owner fails to share this information or to tell the contractor where he can find it, he has breached the contract and is liable for the impacts any lack of this knowledge imposes on the contractor.

- The owner has a duty to act in good faith and to deal fairly in all matters relating to the contract.

- The owner has a duty not to delay, interfere with or hinder the contractor's performance of the Work under the contract.

- The owner has a duty to provide timely access to the site, within the terms and conditions of the contract and the reasonable expectations of the contractor under the requirements of the contract.

- By issuing the construction documents (which are part of the Contract for Construction), the owner warrants that they are *accurate, complete and buildable; i.e., suitable for their intended purpose.*

- Ambiguities in the contract will be construed *against* the writer/originator of the contract (usually the owner).

- Ambiguities in the construction documents will be construed *in favor* of the contractor (so the owner implicitly warrants that he'll pick up the tab for ambiguities that impact the job).

- The owner (including all his representatives) is required to render timely and complete decisions and responses when the contractor raises questions related to the prosecution of the Work under the contract.

- When the owner directs the contractor to perform work beyond the scope of the contract, or unknown conditions impose similar requirements on the contractor, and the owner refuses to execute a change order, a *constructive* change order may be construed to have been committed by the owner to the contractor.

- When the owner is given reasonable notice that the contractor has been delayed on controlling ("critical") work by conditions beyond his control, he has an obligation to modify the contract duration, or the owner will be deemed to have authorized a potential change order to the contractor for "constructive acceleration."

Constructive Change Order

Conditions may occur on a construction project that would warrant the authorization of a change order to the Contract, but for which the owner refuses to authorize a change. Subject to proof that the condition impacted the contractor's performance, the law may recognize that the contractor is entitled to recover damages as though authorized by change order.

Since the contractor's first impression of the owner is through the lens of the construction documents, that's where the seeds of potential confrontation between the owner and the contractor are first sown.

When the documents don't deliver on these warranties, the results are just about inevitable—delays, disruptions, change orders, cost overruns, disputes and claims.

Behind the owner's warranties is the effectiveness with which they're administered. Inept, inexperienced, arbitrary or naïve administration of the contract documents will certainly lead to confrontation.

That is where paradox becomes irony.

When the contractor confronts an owner with assertions that the construction documents aren't buildable or that they're being improperly administered, the owner has two principal options.

He can negotiate reasonably, trying to insert some objectivity into the arguments.

Or he can rely, without independent evaluation, on the design consultants' assertions that the contractor is wrong.

More often than not the owner elects the second option because his representatives—his architect and his construction manager—recommend it (often for their own reasons). It appears (at least temporarily) that *denial* is the best response for his budget.

5

Why should the owner doubt his representatives? He hired them to protect him. It's not comfortable to believe your architect misled you, or that the construction manager's an idiot.

Even though it is at least possible that both of them are in the wrong.

What could be more ironic—or more difficult for an owner to accept—than the possibility that the contractor is the first party he's hired on the project who gave him an honest appraisal of the construction documents?

On the other hand, it can be the owner's opportunism that motivates him to accept this misguided misdirection. After all, the owner knows he's holding the money and may believe he can get away without paying if he applies enough pressure.

Of course, it's possible that all three of the strangers the owner has hired—the architect, the construction manager *and* the contractor—have contributed to the disputes and the mistrust. In this case the owner will have little hope that his budget will come out of a dispute unscathed.

None of these participants told the owner beforehand about the "BEWARE!" sign on the documents when he bought them—or how badly they could bite.

Caveat Emptor

It's likely that every owner has heard the phrase, but never applied it to his construction project. Most of us have the idea that *caveat emptor* means: *"Let the buyer beware!"*

Black's Law Dictionary gives it more weight:

> **Caveat Emptor:** *a purchaser must examine, judge and test for himself.*

This definition is followed by another that's particularly apt for construction documents, though this wasn't the usage originally intended:

> **Caveat emptor, qui ignorare non-debuit quod jus alienum emit.** *Let the buyer beware; for he ought not to be ignorant of what they are when he buys the rights of another.*

You bought the construction documents, owner. You issued them. You bought the consequences right along with 'em.

With the construction documents, the owner buys the rights and duties of all the participants identified in those documents: the owner, the architect, the contractor, their subcontractors, consultants and vendors.

Few architects, attorneys or construction managers advise the owner about those rights and duties before they're committed, bought and signed-off. So it's useful to expand the advice found in Black's specifically to an owner's construction project:

The owner concerned with protecting a project's resources owes it to his family, firm, organization, stockholders, district, agency or taxpayers to *examine, judge and test the contents of those contract documents (agreement, specifications, drawings and other relevant documents) for himself. And to set up the procedures to accomplish that before executing the design contract.*

Most Owners Don't Know What's Coming

Most owners learn only after they've been confronted with a construction claim that they've made the guarantees on pages 46 - 47.

Or that those guarantees took legal effect the instant the owner signed the construction agreement[1].

Since the construction documents define the owner's project for everyone who will work on it, they profoundly affect the coordination, the efficiency and the costs of the work each of them will perform.

They also define the potential for success of the owner's project, or for sinking the owner's budget, schedule and resources.

Despite these realities many owners—some of them large private firms, government agencies, and organizations with continuous building programs—have in place no effective procedures with which to measure the *completeness, accuracy and buildability* of the construction documents before they retain their design consultants.

[1] *The warranties listed in this chapter are not necessarily an exhaustive listing of the warranties that an owner makes when he executes a construction contract.*

5

But The Architect— and His Attorney—Will Tell You...

If the matter becomes a contest, the architect, his attorney and his E&O carrier will tell you that the contractor has an obligation under the terms of the General Conditions of the AIA construction agreement to bring all design errors and conflicts to the architect's attention. (Some of the specific articles they will quote are in chapter 18.)

The architect would have you believe that in the three or four weeks the successful contractor had the documents for bidding he assumed the responsibility to:

- Identify and bring to the architect's attention the errors the architect and his consultants built into the drawings and specifications over the prior eight months;

- Incorporate the corrective work into his estimate and schedule; and

- Incorporate the cost to correct them into his proposal.

While you're digesting this, you might check whether the Brooklyn Bridge is still for sale.

But the Contractor Provides Warranties to the Owner, Too

Whether or not it's spelled out in the contract documents, the contractor is required to:

- Act in good faith and to meet the same standard of fair dealing as the owner;

- Mitigate damages when problems arise;

- Perform all work within a recognized standard of care for such work;

- Schedule and coordinate the work of all subcontractors and suppliers and manage their performance to meet the requirements of the contract documents; and

- Meet a considerable number of performance standards in the various construction disciplines, aggregated under the general heading of "standards of the industry."

E&O—Errors and Omissions

The term usually refers to conditions in the construction documents resulting from the design team's failure to produce complete and accurate documents. E&O insurance is the coverage the owner may require to protect himself against the costs of such design deficiencies in the event they result in recoverable damages by a contractor. (See chapter 12.)

50

The ultimate paradox is for the owner to learn about these two-way warranties from "other strangers" only after he's had to run the gauntlet of a construction claim.

An Early Warning System

There is no "normal sequence of events" that results in a construction dispute.

Because most of the events and actions leading up to a dispute are out of the owner's line of sight, though, it's useful to suggest some typical events, interactions and documents that can lead in that direction.

The events, and the documents that record them, are processed initially between parties who, themselves, can be the causes of a problem. This fact points up the need for the owner to have an early warning system. It demonstrates the importance for the owner to have an objective representative who will receive and review relevant documents.

It also points up the concern that any party that contributes to the documentation process—which includes the CM—can be motivated to hide its own failures of performance.

This fact, together with the need for the owner to develop early, objective perspectives of what can happen during design and construction, is one reason why the owner-guide relationship described in chapter 22 is so important.

Any one of the following actions (plus others) can signal that the owner needs to get on top of events early to avoid confrontations on the project:

- The architect issues an ASI. (See box on page 40.) The architect uses it to "clarify" a condition in the design documents, but is really trying to pass off a change in scope without additional compensation or extension of contract time to the contractor.

- The contractor raises an issue in a progress meeting, requesting information. The successive meeting reports show that the matter is still unresolved several meetings later.

- The contractor submits shop drawings that the architect returns without accepting them. (Design consultants no longer "approve" anything the contractor submits, on the

51

instructions of their E&O carriers.) The architect's "rejection" may require that they be corrected or modified and resubmitted. Or they may be accepted with specific instructions to modify them. Sometimes the unscrupulous architect will attempt to return submittals with built-in design changes, in the hope that he can slip them by the contractor without the need for a change order. (The scruples of the architect can be a direct reflection of the owner's scruples, too.) The timeliness with which these submittals are initially transmitted and subsequently processed can have major implications for the project. The submittal process is of critical concern to the owner. The architect and contractor are jointly responsible for ensuring that the process is consistent with the requirements of whatever has been accepted as the controlling schedule for the project.

- The contractor forwards a Request for Information (RFI) to the architect. Whether the answer the architect provides is timely or delayed, responsive or inadequate can have long-term implications. The issue raised can be spurious (some contractors raise dust just to cloud the job environment or to hide their self-caused problems). The RFI can result in a simple clarification by the architect, with no cost or time implications. Or it can start a continuing series of events that culminate in a change order or a dispute.

- The contractor forwards an Extra Work Order (EWO) to the architect. This is essentially a notice that the contractor has identified a condition that he believes will require a change order. It's a notice that the contractor has bypassed the RFI procedure and already drawn his own conclusions about the need for a change order. As with the RFI, however, some contractors will open an EWO file on their books, accumulating a record of insubstantial, specious complaints just to create a large record, in the prospect that it will lend weight to a future claim.

- The project schedule is a pivotal submittal document. It needs to be sufficiently detailed for all the parties, including the owner, CM and architect to know how they are to relate to each other in order to deliver the project within its contract duration. It needs to include the scheduling of:

1. The interactions between trades;
2. The interactions between submittals, deliveries of materials and equipment, and field work;
3. The interactions between what the owner has required the contractor to do, and the promised or implied actions of the owner and architect.

This standard is seldom met. Contractors fail to meet it in the preparation of their schedules, and often disregard it after the schedule is submitted. Design consultants, and more construction managers than owners realize, are not competent either in their understanding of construction logistics or with definitive, effective scheduling techniques.

The measurement of how the use of time is planned and the documentation of its actual use is of such importance, however, that all of Part 6 is committed to a discussion of *time management.*

Since most of these events are conducted behind closed doors and are initially invisible, how does the owner get an early warning?

The answer needs to be: through delivery to the owner, concurrent with their regular recordation, of all the logs and documents that record these transactions.

The suggestion *BIT* makes to the owner is that he or she incorporate into the Project Plan procedures whereby meeting notes are monitored and reviewed by someone without an axe to grind—someone not party to one of the performance contacts (architect, contractor or CM); and that copies of the logs of all transactions between all parties—daily reports, transmittals and correspondence—are regularly transmitted to the owner.

The Proof Is In the Documentation

Any dispute between owner and contractor that results from testing one set of warranties against another requires that the asserting party prove its case. This process is called in the law *proving the merits of the claim.*

Proving—or disproving—requires documentation. Testimony may eventually bear some weight, but the solid arguments will come from records maintained concurrently with the prosecution of the Work under the contract.

5

The central document will always be the Contract. The Contract itself lists its components: essentially all the documents the contractor requires to build the project. The owner's first responsibility when considering construction is to understand clearly what is contained in the Contract. He needs also to understand that, whether or not they are spelled out in the contract, there are unwritten—"implied"—warranties of performance on which he, his design consultants, and the contractor must deliver. These are called generally "standards of performance." They have been established by the customs of the professions and of the construction industry.

Plus the standards that have been established by precedent under the Law of Contracts.

Plus applicable state regulations and codes.

How many parties sign a performance contract with more than the merest understanding of these requirements and warranties? The exponential increase of construction disputes over the past quarter century is a clear answer to that question.

Then, there is an almost endless stream of project-generated documents, any of which can have some bearing on the resolution of a claim. This is an abbreviated list: specifications, daily job reports, meeting minutes, RFIs, change proposals, submittals, EWOs, ASIs, change orders, plus the associated logs that go with all of these documents, plus payment records, schedules, cost reports, timecards, conformed drawings, as-built records, correspondence, transmittals, delivery tags, lien releases, and many others.

The contractor has all or most of them. Creative attorneys and "claims experts" retained by the contractor can evolve very convincing cases out of them.

Most standard construction contracts leave the owner naked. These records are often kept in the archives of strangers who may have their own territories to protect when the project is completed.

Few contracts guarantee that the owner will have his own equivalent, objectively maintained copies of these records.

This is an area the owner learns of only after things begin to fly out of the fan. These documents are difficult to obtain from *any* of the parties once a claim has been filed. It's useless, as we'll see in chapter 13, to rely on some statement in a contract that the owner has a right to demand production of documents once a dispute arises.

There are too many ways documents can be "lost," modified or re-invented. A surprising number of fires and thefts have occurred in project jobshacks. Attorneys have more ways than most owners can imagine to block their production until access to them will bring the owner no benefit.

Even if they are "produced" or "discovered" after a claim is initiated, the owner will lose the considerable advantage of momentum in his defense of a claim if he's seeing them for the first time, and if he must then hire "other strangers" to help organize, review and interpret them.

All of these concerns raise issues of the usefulness of all these documents, and of the reasonableness of providing space for them.

> *Will the owner actually need these documents? Will it ever be necessary to review them? Where can they be stored? How can they be organized? If they're never needed, won't the owner just be wasting effort and money?*

It's my experience—both as owner's representative and as one of those experts from the claims-resolution cottage industry—that even though they may never be required, the cost of having all parties deliver the documents that record the interactions on the project will not materially add to the cost of either performance contract.

It's my experience that these documents can be a valuable resource not only in the event of a dispute, but for effective negotiation of changes in the course of construction.

The owner about to undertake a building project needs to consider before he hires the architect what he will put into both contracts—the architect's and the contractor's—regarding delivery of documents to his archives in the course of their work. He also needs to consider how he will make use of them.

An alternative is for all parties to deliver to a single location, accessible at all times to the owner, selective documents the owner believes he will need to keep abreast of concurrent activities. This facility should have copy equipment so the owner can make and take away copies for his own files if necessary.

The City Hall

Typical of the stories you will read in this book, some of the facts here became evident to all of the participants only after the project had experienced a claim and had gone to arbitration.

The City retained an architect with a reputation for similar projects to design its $1.8 million city hall. It also retained a construction manager to oversee the project and to provide value-engineering services. The City instructed the CM that *no change orders would be allowed.*

Despite the CM's "value-engineering," the construction documents that went to public bid for lump-sum quotations were incomplete in many details. The status of the several design disciplines was somewhere between 75% Design Development and 35% Working Drawings.

The building had different angles and shapes, joined in planes that were not readily apparent. Instead of showing these interfaces, the architect had drawn "sections" through straightforward locations and avoided the difficult areas.

As the contractor started developing field layouts and shop drawings, it became evident that the dimensions of the foundations and the building frame did not match, that the dimensions of the building framing itself did not "close," and that the roof layout did not match the framing.

The work required redesign on a continuing basis. The contractor was directed to *"work it out through shop drawings"—was told that it was a contractor coordination problem.* The contractor requested compensation for extra work, additional materials, delays and loss of efficiency.

Despite all the evidence, the owner, architect and CM went into denial mode. The contractor was directed to accelerate his work to make up for delays.

The contractor ultimately won his $350,000 lawsuit, after all parties had spent over $202,000 in legal fees. Only the mediator and attorneys won.

There is another subtlety to the concurrent sequestering of documents by the owner. Documents produced in the course of ongoing work are much more likely to be real and not subject to denial. If arguments have not yet started, they're less likely to contain self-serving information.

The existence of these documents, and the need to use them effectively suggests that the owner may need to have available someone on whose judgment he can rely to interpret them. This resource—someone I've called the guide—is an owner-resource we'll discuss in detail once we've visited the territories of the strangers the owner will retain to develop his project. One of the guide's assignments may be to assist the owner to establish a system to organize and access these documents once they start to arrive.

The Contractor Can Be a Valuable Resource

Despite the caveats and reminders that the owner needs to look for all the leverage he has available to maintain objectivity on the project, it's important to reprise the theme of *respect*. Respect will flow to the owner who produces a clear set of directions—design documents and administrative system—with which the contractor can produce a successful project.

Either in the course of doing that, or even after those documents are published, the owner who has generated the potential for that respect will find that a contractor can, himself, contribute considerable creativity to a project's execution.

Hardly a copy of the *Engineering News Record*[2] *(ENR)* has arrived in my office over the past 35 years that didn't introduce some inventive approach a contractor devised to solve a tricky design or construction problem—to reduce irreducible time, to propose a better system, to span an unspannable space, or to engineer a one-of-a-kind piece of support equipment.

In Chapter 19 we'll discuss some of the contractual options an owner has for involving the contractor early. Some of them will make the contractor or "construction manager/contractor" the lead entity on the project.

[2] *ENR: one of the signal magazines of the construction industry, and a good resource for an owner with a continuing building program.*

5

It's important, however, for *every* owner to establish and maintain a consistent view of the construction-contracting relationship. Many projects fail because the owner goes into a CM or design/build relationship with the same pie-in-the-sky optimism with which others retain an architect.

Unless the owner establishes a check-and-balance system for measuring and documenting *all* performances, that optimism can turn, as it has in more projects than can be counted, to disillusionment and to destruction of the owner's budget.

18% savings in Costs—44% savings in time

The bank had built branches before, usually taking about 8 months for each branch. It was about to undertake a major branch-expansion program.

JBA was retained to develop and implement new procedures on the first three projects. All three were completed within budget, and delivered in 6-1/2 months or less.

JBA's assignment was expanded with four additional projects. By the time these four were completed, construction time was down to 5-1/2 months. Costs had been reduced 9%. The program continued over several years and JBA ultimately assisted the bank with the planning, design and construction of 40 branches.

By completion of the 10th bank, construction durations were down to 4-1/2 months and average costs had been reduced by 18%. These savings continued through the program.

These results were achieved even though each bank had a different architect and contractor, and was of a unique design, appropriate to its community. The results came from: (1) standardization of administrative procedures for design and construction; (2) elimination of design details that resulted in delayed deliveries; and (3) pre-construction assessment of specifically local construction logistics.

3

FOREIGN TERRITORIES

For most owners, the territories in which designers, contractors and attorneys work are like foreign countries, each with its unique language.

The next several chapters will introduce you to those territories and add new vocabularies to the one we started to pick up from the contractors.

6

STRANGE VOICES FROM
THE BLACK HOLE

Litigation, like a galactic black hole, draws all sorts of strange bodies into it.

You're the owner of the project. Your contractor has brought a claim against you for delay, disruption, loss of productivity and the costs associated with asserted design defects and administrative deficiencies.

On the advice of your attorney, you've brought your architect into the matter as a witness to assist in responding to the contractor and his subs. Your architect is not yet a party to the claim because you don't want to initiate a premature divorce and lose the support of your design consultants. But who knows? You may find yourself at odds with him, too, before it's over.

Nobody gets a chance to tell a simple, straightforward account of what happened. Everything has to be brought out through a tedious Q&A process.

You've been through months of complaints, cross-complaints, interrogatories, document production, audits, discovery and depositions. Despite all the paper that's flown back and forth (most of it buried in unintelligible legalese) asserting that you were not responsible for changes, you've already spent $37,000.

You're in the courtroom waiting for the bailiff to announce, "All rise." You look around, still wondering why you're here, and realize how different your reactions are from watching court proceedings on TV. It's one thing to see a strange country on TV. It's another to experience the sounds, the smells, the anxiety and stress of actually being there.

The first things you notice in a foreign country are the differences in language. Not only do some of the inhabitants use unfamiliar words and phrases, but your own language also sounds different when some of the natives and other strangers there speak it. The attorneys use a jargon only they understand. Nobody gets a chance to tell a simple, straightforward account of what happened. *Perry Mason* or *Matlock,* it's not.

61

Everything has to be brought out through a tedious process of questions and answers. The project you're chomping at the bit to tell everybody about from your own viewpoint, comes out instead in small sound bites from multiple witnesses. Some of these faces are familiar; some are complete strangers. Long pauses, recesses, lies, "objections," "motions," "rulings" and multicolored tables and charts interrupt their stories. The project you thought you knew is distorted beyond recognition.

Then, you start to realize it's not just differences in language. It's not just the confusing charts. It's not just the disruptions and delays that have distorted the history of your project.

It's that the construction documents that represent your project have delivered distinctly different messages to different people, natives and visitors alike.

The contractor takes the witness stand and describes how he was delayed and impacted:

> "The construction documents weren't _coordinated._ We couldn't build from 'em."

The contractor's attorney puts your witness, the architect, on the stand and asks him:

> "Is it true that you received 324 RFIs from the contractor?"

Your architect responds:

> "Most of those RFIs were meaningless. The contractor failed to _coordinate_ the work of his subs."

The contractor's attorney turns to the Court:

> "Objection. Move to strike on the basis of being non-responsive."

The Court rules:

> "Sustained."

The contractor's attorney puts the electrical subcontractor on the stand and asks:

> "Were the construction documents buildable?"

Your attorney rises:

"Objection. Vague and ambiguous as to 'buildable.'"

To which the judge says:

"That's all right. We've already heard about 'buildability.' I'll allow it. Overruled."

Despite the fact that, as your attorney will tell you, the question should not have been allowed in any case since a percipient witness cannot express an opinion in his testimony—but the judge rules.

And the contractor's attorney says to the electrical subcontractor:

"You may answer."

And the electrical subcontractor replies:

"No. The electrical underground and the risers weren't coordinated with the layout of the structure."

Several days and a dozen witnesses later, your architect returns to the stand on redirect examination by your attorney, who asks:

"Did the contractor perform within the requirements of the contract?"

Your architect responds:

"No. He was completely incapable of coordinating the layouts and the shop drawings."

And later, somebody's expert (yours, his, theirs...) asserts:

"The schedule demonstrates why the job was uncoordinated."

Over a few days the word *coordinate* has been used to criticize your construction documents, to accuse your architect of being non-responsive, to attack the contractor's performance, and to put fault on the project schedule. All you know is that *somebody's* inadequate coordination didn't compensate for *somebody else's* failure to coordinate *something or somebody.*

6

One thing you can be sure of: you'll spend a whole lot more money before the confusions surrounding *coordination* are sorted out—if ever.

At this late date, one more thought occurs to you:

> *"I wish I'd studied the documents and the contracts a lot closer before I signed 'em."*

Strange Voices

As you listen to the strangers on your project you realize that not only do they use buzzwords and phrases unique to their professions and trades:

> "CAD," "specs," "boiler plate," "RFIs," "bundling," "rebar," "standard of care," "T&M," "learning curve," "force account," "entitlement," "Eichleay formula," "motion to compel," "res judicata," "stipulate to..." "doctrine of laches," "motion in limine," "retraxit," "duces tecum," "quantum meruit...."

They also use ordinary words in ways that have unique meanings for each of the persons using them:

> "impact," "coordinate," "accurate," "promptly," "discover," "acceleration," "periodic observation," "interrogatories."

The dialogue—especially in the environment of a construction claim—often seems less concerned with extracting "truth" than obfuscating it.

It is one of the purposes of this book to assist the reader to become at least comfortable enough with the vocabularies of design, construction and the legal process that surrounds disputes and claims to challenge the concepts and panaceas that would otherwise leave you in doubt about what's being done in the name of your project.

Strangers in a Foreign Country

The hours grind on. Arbitration and trial can be both really disturbing and really boring. The interruptions, motions, conferences in chambers, recesses, continuances and delays completely destroy the possibility of presenting any reasonably intelligible project history.

Gradually, another realization starts to get through to you: most of the inhabitants in your landscape—there are a lot more than you realized—are strangers to each other too. Each has his or her unique viewpoint. This puts not only the court-room, but your own project into the category of a foreign country for most of them, too.

Some of them have been to places like yours before; some of them even worked together before, in different places. But this is the first they've come together in your documents, at this time, in this landscape, with these contracts, this architect, this manager and for this owner.

Most of them came without knowing how the relationships—especially the ones with this owner and this architect—would work out. They're weren't sure of each other then, and they still aren't. They're wary.

It would have been unusual if they were entirely comfortable with each other. Each was driven by the need to make a profit on *this* job. Relationships were formed (if only temporarily) in terms of where the cooperation to realize that profit could most likely be found.

You realize that you're here because your design consultants failed to provide clear and accurate directions to these strangers. You learn that neither your specifications nor your construction man-ager were up to the job of solving the problems those documents created.

You begin to understand just how many strangers were invited to set up camp in your construction documents. And as though the strangers who've already been through your landscape weren't enough, there are all those other people in court you didn't even know were out there all along, waiting.

The sureties and their attorneys, the "experts," the auditors—and all the other parties' attorneys.

In order to get a handle on who they are and how they got together with you in this courtroom, we need to go back to where the process started that brought them, and you, here.

To do that, we first need to define the phases of project development.

The Self-Education of School Boards

The state of education in America is frequently attributed to our cultural and teaching standards.

But it can also be charged to the wasting policies that state departments of education have been lobbied to include in their procedures for the administration of school construction funds. The ways and means of this, as with most political boondoggles, are buried in the labyrinths of politics.

Unlike the untraceable disappearance of other funds, however, some of this waste is potentially reversible by individual school administrations. A knowledgeable school board can control how the design and construction components of its funds are budgeted, directed and managed.

How are they wasted? Consider:

- Many schools have ancient infrastructures: underground utilities, distribution systems and exterior surfaces that need repair or replacement. 30%, 50% or more of the construction funds must often go into these systems before improvements can be made to other facilities. Departments of education have been lobbied to suggest through their regulations that the design for *everything* in a school program must be done under the aegis of an architect, and to suggest architectural fees in the order of 9% to 12% for renovation work. But the architect will do little more than put his stamp on the utilities and distribution system drawings an engineer will design for a fee of 3% to 5%. These state guidelines, however, allow the knowledgeable district to carve this work out of the architect's agreement. (A 5% saving on $8 million comes to $400,000.)

- Many school districts use county-provided contracts. But it's within the prerogatives of most districts to formulate their own contracts and to require the responsive services discussed in this book.

- Then there is the *Mission Statement*. School administrations, often notable examples of the Peter Principle when it comes to construction, lead their boards and committees to believe that a directive to an architect *To Keep Our Students Warm, Safe and Dry* is an adequate program description. With a multi-million dollar modernization program in the offing, this is just a license for the architect to run free with his *muse* and with the district's funds.

Measured in terms of school districts that run a tight ship, a school district can save 5% and more of the consulting funds hard won from local and state bonds.

7
MAPPING THE TERRITORIES

An owner who has been through project development before will immediately recognize that there are many more events and influences on a project than we'll cover with the following descriptions of these basic phases.

On many projects, it may be necessary to undertake extensive site explorations to evaluate existing conditions. Intensive environmental studies, and even separate remedial contracts, may have to be undertaken before the development of a project can even be considered.

There may be need for extensive exploration and survey of existing structures and dimensions. Destructive testing and exploration behind existing surfaces may be essential before design can start on a renovation project.

We won't cover these special conditions here. If they do influence a respective project, they will be among the additional considerations the owner needs to incorporate into his or her Project Program and Plan.

Our purpose here is to cover the basic, essential phases that are common to every project. Experience has demonstrated that, as universal as these are, many owners don't understand what needs to happen in each of these phases to produce a successful project.

You'll find that if you can get through these basic phases efficiently you'll be far ahead of owners who end up being confronted with disputes during construction. If you can optimize the use of the resources *BIT* recommends during these phases, you'll have gone a long way toward resolving those other special concerns at the same time.

Essential Interactivity

So many things have to be planned, scheduled, managed, shifted, caught up with, funded, recorded, measured, documented, changed, and changed again, in the course of design and construction that the lines between the phases we're about to define

7

can become blurred. This is especially true for the owner who's occupied with other professional and personal demands[1].

But it's essential for the owner of a prospective project, in order to get a handle on how the performances of all the participants need to be coordinated, to start with a picture of the basic scopes of work in their service agreements.

Those scopes of work are grouped into seven sequential phases. Because the lines between these phases can be blurred, and even erased, it's not necessary to think of these phases as having hard and fast definitions. Some of these—design, for instance—have subphases we'll discuss in chapter 8. Even the description of what goes on in each phase can be different depending upon whose buttons are being pushed.

We'll visit these phases several times in any case, each time with a different perspective. We'll visit them as we consider how to measure and document each phase. We'll discuss them again as we review how the standard agreements promoted by various professional associations need to be modified in order to obtain responsive performances.

By then, the standard definitions won't matter. You will have developed—in your notebook—your own ideas of the procedures and service agreements you'll need in order to manage your project. You'll be prepared to require specific scopes of work and specific performances for each phase. For this purpose I've added an eighth phase: Project Management.

The basic phases described generally below include:

1. **Planning and The Project Plan**

2. **Programming**

3. **Funding**

4. **Design**

5. **Bidding and/or Negotiating the Construction Contract**

6. **Construction**

7. **Post-Construction**

8. **Project Management**

---

[1] *This preoccupation with other activities is one of the reasons for the suggestion that an owner find a guide early in the development process—before hiring any of the parties who will be producing it.*

1. Planning and The Project Plan

Planning and *Programming* are sometimes used interchangeably.

This demonstrates one of the problems of construction vocabularies. The same words are used to mean different things, while some identical ideas are often called by different names.

Planning is often associated with *Master Planning,* a term with its own special meaning. Master Planning usually refers to the overall concept of a project. For example, the overall layout of a cityscape, subdivision or high rise and its relationship to the surrounding environment. Master Planning may or may not incorporate all of the elements of *Programming (see 2 below).*

We'll use *Planning* in relation to the central purpose of this book: your development of a Project Plan that defines **how** your project will be managed, measured and documented.

2. Programming

Your *Program,* on the other hand, will define **what** you expect to have when your project is complete. This is consistent with the typical use of the word Program in the industry.

Programming is the process through which the owner initially defines what he needs and wants (preliminary planning). It's also the process he continues with his consultants, refining these preliminaries into what the owner's resources can produce, and defining the criteria that will drive the development of the construction documents. The development of a budget is an essential element of Programming.

The Program will define the functions, the operational requirements, the circulation patterns throughout the project, the standards for quality, materials, aesthetic values, operations, maintenance, levels of staffing, services and systems to be built into the project.

To be complete and effective, the Program also needs to incorporate a project schedule that defines clear time limits for design, construction and occupancy.

The levels of effort and programming detail required to advance the schedule and the budget through completed construction documents will be very different for a private residence, detention facility, library, vacation home, medical

7

lab, university, condominium, shopping center, manufacturing plant, hospital or airport.

Some programs will be several pages in length. Some will fill volumes.

It's essential that, whatever the program's content, it be in writing and that it is agreed to by all parties—signed off—before the start of design. It may be modified as things go along, so this is a critical reason to start with a confirmed, clear statement of the owner's needs or wants.

The authorized Program is an essential element of the agreement between the owner and the design consultants.

3. Funding

While thinking about the checks and balances that will go into the Project Plan, an owner needs to realize that project funding and the construction budget intersect right in the heart of the architect's design contract.

At the end of programming and every phase of design, the architect is required to provide the owner with a "probable" or "estimated" construction cost of the project.

What "probable" cost means, who prepared the cost analysis, what parameters were used to develop it, who checked it and how it reflects construction realties will all have major consequences when the project documents are delivered to contractors.

The standard AIA agreements (and others like them) lack credible definition of what "probable (or estimated) construction cost" means[2].

Usually, no one knows with certainty during the programming and design process exactly what the project will cost. Even successful projects vary from their owner's hoped-for budgets.

No matter how closely the final "estimated cost" and "construction contract duration" that come with the completion of construction documents compare with the owner's original ideas about costs and time, they had better be close to what

[2] *In fact, the commentaries provided to architects as instruction on how to use the standard AIA documents suggest that the "estimated cost" be a loosely interpreted description (see chapter 18). Anxiety about liabilities, or ineptitude?*

the contractor will find in those documents if the owner expects to realize a successful project.

4. Design

Design is the aggregate of all the efforts required to translate what's in the owner's preliminary program into construction documents (CDs).

The subphases into which the design consultants' efforts are divided to produce these documents are usually:

- Programming,
- Schematic (or Conceptual) Design,
- Design Development,
- Construction Documents (sometimes "Working Drawings").

These phases are listed but not defined in the standard AIA agreement (and other similar standard forms). The agreement provides for payments to the architect for each phase of design, usually in terms of meaningless percentages: "75% design development," or "90% construction documents."

These standard agreements do not contain adequate descriptions of what these successive phases will produce. They do not have useful definitions of the design consultants' scopes of work, the scheduling of their performances, nor the means to measure and document the adequacy and timely delivery of the design products promised.

We'll discuss the development of meaningful definitions of design services in chapter 9. For now, it's enough to report that few owners have ever seen design documents that matched the percentages of progress claimed for them.

Addendum

This is a modification to the construction documents (possibly including the specifications and bidding documents) issued after the CDs are issued, but before bids are received. Contractors are required to incorporate these modifications into their quotations.

5. Bidding and/or Negotiating the Construction Contract

The standard design agreements overlook a lot that happens during the bidding period.

The way in which the owner's representatives administer the bid documents can have major implications, however, for how contractors will preview the project. It can have major implications for events during construction.

7

The issuance of addenda during a bid period can signal future problems on a project. The fact that addenda are even required raises questions about how the owner and architect have handled the planning and design phases.

What the addenda contain, the changes they make, and the manner in which the addenda are administered during the bidding period can telegraph some potently negative messages to contractors.

Pre-bid and pre-construction meetings need to be carefully set up, managed and documented to avoid asserted disputes: *"What we heard the architect say in the pre-bid meeting was...."* The standard design agreement leaves the management of these meetings entirely to chance.

The Carrot

During the bid period for the budgeted $11 million manufacturing project, the architect revised and reissued 75% of the drawings as part of an addendum during the 3-1/2 week bid period. The bid period was extended one week. After the contract was awarded, but before it was executed by the owner, the architect again revised and reissued 75% of the drawings.

The project had been bid with the expectation that [1] if the contractor performed well, and [2] the owner proceeded with its program, the contractor would be awarded a $5 million change order for Stage Two. Despite the impact of the changes, the contractor knuckled under quietly and attempted to incorporate the architect's changes without raising dust.

The contractor continued to be benignly cooperative when the only way to install foundation rebar was to burn away several heavy bars in each footing (which the architect authorized without blinking). The contractor also quietly accepted the impacts when it found that many elements (for example: mechanical ducts, electrical panels and metal stairs) were all in the same location on different drawings and had to be field-relocated in order to be installed. (The architect asserted they were "field coordination problems.")

But the contractor woke to reality when the owner announced—halfway through the project—that it would not be proceeding with Stage Two.

The contractor brought a $1.7 million claim to recover its costs. The matter was adjudicated in a heinously long (four-month) arbitration (with one arbitrator half-asleep through much of it). Combined costs to all parties ran to $780,000. The contractor was awarded $1.7 million. Only the attorneys, the arbitrators and several consultants came out whole.

In order to measure the responsiveness of contractors' bids, the owner needs carefully defined criteria before the documents go out to bid. This is another area notable for its absence in standard design-services agreements.

The bid documents and the procedures for administering them are unique to each project. Their definitions will be determined by the characteristics of the project and by the type of construction contract to be executed.

The owner planning a successful project needs to invest considerable thought into these procedures while the CDs are being produced, well ahead of their issuance.

6. Construction

In the chapters on contractors, we have already discussed how contractors perceive a project and how these perceptions are the products of everything that happens in the planning, design and bidding phases.

The owner's decisions regarding how design will be managed will not only enhance the owner's influence on construction performances, they will provide the owner with insights that few owners have about their own projects.

7. Post-Construction

In the course of construction, the contractor should have been required to keep the construction documents and specifications marked up with any changes and confirmations of location that evolved during construction.

In order to ensure this effort, and the accuracy of the resulting markups, someone needs to check these markups monthly on the owner's behalf as a precondition of payment to the contractor.

The architect should be required by his contract to incorporate these changes at the end of the project, and within a specified time, into a reproducible set of *conformed construction documents, or Record Documents (sometimes referred to as "As-Built Documents")*.

Even in good drawings, the work of some disciplines will be shown as a single line. The single lines that represent the location of electrical conduit and plumbing distribution are examples. Because of their inherent flexibility their final layout is left for the contractor to fit these systems into available spaces.

Once these systems are installed in their final locations, it's extremely important to the owner's possession, operating costs, maintenance and future use of the project for these to be shown in their final locations on a modified ("conformed") set of the construction documents. The same is true of any systems, materials or equipment that required changes in the course of construction.

Conformed Drawings and Markups

Construction drawings that are "conformed" have been updated in the course of construction to incorporate actual conditions on the project. The information for these conformed drawings is generated by having the contractor "mark up" on a regular basis the actual locations of, and/ or changes to, work installed in the field.

When materials or equipment have been modified, it's equally critical to conform the specifications to reflect their "as-built" conditions.

If the project is long and complex, the owner might elect to have *conformed drawings* updated by the architect at intermediate periods during construction, as well. This may be the only way to assure accuracy, and to assist the contractor by providing drawings his subs can reasonably work with.

It should be the contractor's duty to keep up the drawings with all addenda, changes and final locations of installations. The contractor can be committed to this duty by his contract. This accuracy is of paramount importance to the owner, who needs to establish some independent means of checking these markups. The inspector can become a resource in this regard, depending upon his relationship to the owner.

The competence and the accuracy of these documents are of sufficient concern to the owner that he should include in his budget a cost to ensure that these sets of documents are competently maintained as work progresses.

These accurate As-Built Documents of the project will be an invaluable resource for the owner for future renovation and to assist with the operations and maintenance programs on his project.

This is one more reason for the owner to invest long-term thinking into the development of the Project Plan.

Provision must be made in one or more contracts [1] for someone to confirm in the course of construction the accuracy of the markups that record these final locations and changes; and [2] for some party to transfer this information into a finally conformed set of construction documents (the "record documents"). The usual party to do this is the architect. If this is the case, the standards for accuracy and timeliness of these conformed drawings needs to be spelled out in the architect's agreement.

Failure to produce an accurate and complete set of reproducible record documents can cause the owner considerable agonies and extra costs in the future.

8. Project Management

Project Management is not specifically identified in the standard agreements produced by the associations identified in earlier chapters.

But management of an owner's project is implicit either by reference, or by oversight, in every article of every one of these agreements.

Management is the central subject of the Project Plan you'll be considering once we return from this trip.

This is a good first place, therefore, to consider the perspectives you will need to assemble the jigsaw puzzle represented by those seven phases above.

MANAGEMENT

Management: A Balance of Experience and Art

The dictionary says that managing means:

> *To bring about; to succeed in accomplishing; to dominate or influence by tact, address or artifice; to handle, direct, govern or control in action or use; to bring about, usually despite hardship or difficulty.*

On a construction project, this *directing* needs to be *balanced* with the interests of those strangers.

Otherwise, it will become a contest, and there's no question that the owner will be one of the losers if a contest occurs. There are too many strangers out there doing too many different things for it to be any other way.

The management challenge is to integrate a project's resources into a cohesive plan.

Managing those strangers requires a balance of *experience* and *art.*

The *experience* required is the kind that knows what happens behind closed doors during the design process. It recognizes what is required to ensure the buildability and budgetary responsiveness of a set of construction documents. It has competence with the tools required to measure and document design and construction performances. It has used those tools effectively during the bidding process and out in the field during construction. It is aware of what happens in jobshacks, in design offices and in progress meetings when clear guidelines aren't established. It recognizes the differences between useful project documentation and wallpaper.

The *art* of management can be summed up by the word *active*. Management operates above *reactions* to "what's going on."

GMP—Guaranteed Maximum Price

Also: GOP: Guaranteed Outside Price. This concept for defining the upper limit of construction costs is applied to any of several contract formats: Cost-Plus-Fixed Fee, Fast Track, Segregated Prime, etc. The suspect word here is guaranteed. As we all know, everything we buy today has a *limited* guarantee attached to it. Offerers of GMPs always associate limits to their guarantees, too, but are not always up front with what they believe those limits are—until they identify a condition they believe they can use to explain why the costs just passed the limit. It is absolutely essential for an owner to receive definitions of scope, quantities, quality and other limits associated with a GMP before relying upon it.

It requires comprehensive knowledge of the processes being managed. Effective, pro-active management requires detailed knowledge of the contracts and the construction documents and of the processes involved in programming, design, budgeting, scheduling, construction, and productive negotiating.

Effective management does not wait for events to take place: it anticipates them. It first sets up scopes of work and criteria that are required to accomplish the project's goals.

Effective management balances a thorough knowledge of what happens in design offices and in jobshacks with the standards that should be applied to what happens there.

Then, it *actively* measures performances in relation to those standards.

During design, this means tracking the products of the design consultants for *coordination* and *accuracy*.

During construction, it means tracking the performances of *all* the parties on the project (including the owner's participation in decision-making) in relation to the requirements and limits defined in a project schedule and budget.

The art of management includes the ability [1] to articulate clearly the standards and scopes of work to be performed by those being managed, [2] to measure the performance of those scopes, and [3] to use effectively all the leverage available to obtain the owner's goals.

The art of management is to be *proactive* instead of *reactive*.

The Product of Reactive Management

The *art* of management also consists of the ability to represent the owner effectively in a fair and balanced way.

This art requires the self-assurance to listen quietly to people who have differences.

77

7

Even with the best of documents, construction is a confrontational environment. It takes a blend of experience and maturity to keep others' confrontations from getting out of hand.

For all the platitudes you will hear to the contrary, confrontation is implicit in the concept of management: *to confront difficulties, to overcome resistance, reluctance, or just plain indifference.*

So is the ability to negotiate those differences without raising blood-pressure levels or adding to potential conflicts.

> *Is this level of judgment available from the junior engineers, retired inspectors, or the "schedulers" that many "construction management" firms send to the field? Do architects, cost estimators, or construction inspectors, who have spent most of their professional careers in areas other than management, develop these resources overnight?*
>
> *Can an owner obtain this mature and objective representation from the same design consultants whose products are the principal reasons for those confrontations?*
>
> *Can the owner receive objective representation from a contractor whose "construction manager" was the contractor's own assertive representative on the contractor's last project?*
>
> *What impact will it have that the "owner's construction manager" may be dealing with the same subcontractors that are working for the CM-general contractor on other projects?*

By the time you start to develop your own Project Plan, your responses to these questions may be very different from what they were before you started this journey.

4

DESIGN

The initial challenge for the owner is to recognize that design consultants are essentially strangers; that many of them take on responsibilities that exceed their capabilities.

The second challenge for the owner is to define their scopes of work, and measure their performances with the same tools with which contractors' performances will be measured.

8

THE DESIGN CONSULTANTS

Your Project—Your Book

The day you decided that you would develop your building project, you became an author.

The book you intend to write is the one you expect your contractors to follow as they deliver your project to you. It has several sections: the Construction Contract, the Bid Documents, the Administrative Specifications, the Drawings and Details, the Specifications.

The quality, timeliness and cost of your project will depend upon how well your book reads, how well it's been edited (*"coordinated"*—the word we heard in different contexts in court in chapter 6). Your project's success will be a direct reflection of how many errors your book contains, how well it covers the ground (responds to your Program and complies with codes and regulations), and whether it's clear and unambiguous.

You don't feel qualified to write the entire book yourself, so you hire a ghostwriter to work with you to put it all together, to make it readable.

But, despite the fact that you recognize you need help to write it, you're determined that when it's completed your book will represent your ideas, modified only to the extent that your ghostwriters bring value-added creativity to your concepts.

Your Ghostwriters

When your book is published, it will be your name on the cover. Your ghostwriters will be identified under "written with."

Once your project's complete, it will be your name on the mailbox, your logo at the entrance, your new address on your letterhead. You'll be living, working or sending your kids to school in what your book produced. Your design consultants won't be in sight; only the evidence of how well they translated your ideas to contractors. You will be carrying the financing, operating and maintenance costs for the next 20, 30 or 40 years.

81

8

How closely do you want to monitor what your ghostwriters are doing with your ideas? How often do you want to see galley proofs while they're doing it?

You need to answer these questions before you hire them.

You expect your consultants to bring their special talents, creativity and abilities to your goals.

But it's your resources—your story, your money, your financing costs and your time—that will go into your project. You're not making these investments so an architect can realize his *muse*.

Architecture has been described by architects as something of an adventure in *chaos:*

> *"Architecture is a hazardous mixture of omnipotence and impotence. Ostensibly involved in 'shaping' the world... architects depend on the provocations of others... clients, individual or institutional. Therefore... Architecture is by definition a chaotic adventure[1]."*

An effective architect will extract and organize the owner's ideas of what the owner needs or wants. In the process, the architect will expect the owner to "provoke" him into using his creativity.

While the architect is challenging an owner to be clear about what he or she wants, he is, whether the owner knows it or not, challenging the owner to define the limits for that creativity. There can be a chasm the size of the Grand Canyon between the architect's provoked creativity and the owner's management of those limits.

How much chaos can a project stand? How much of the architect's muse can the owner afford to buy? While you're musing about your answer to this question, look at the buildings that surround you. What features of these buildings are purposeful? Which are ornaments? Which are obviously advertisements for the owner; which for the architect? Which are downsized carbon copies of ideas invented 15 cities or a dozen school districts ago, copied a hundred times since[2]?

---

[1] S,M,L,XL, *a "novel about architecture", whose subject is primarily an exploration of how architectural space evolves.*

[2] *From Bauhaus to Our House is an interesting commentary on the proclivity of architects to copy from a limited number of masters of their profession.*

Management and Reality

There's nothing inherently wrong about giving an architect his head. But it's important to know in advance where the limits are. An owner needs to keep close control on his goals, his planned uses and his budget. He needs to know what the *muse* is costing him, and he needs to know when to say *enough already.*

The architect's *muse,* his images of *what can be,* have nothing in common with the resources required to put a construction project onto a track toward reality and keep it there. The initial separation of the two occurs in the Programming and Schematics Phases of design.

Once the architect has translated the owner's ideas into *schemes* about which they both *think* they can agree, the owner needs to impose those budgetary limits and to measure where the balance of the design goes from there.

The AIA documents leave entirely in suspension and without adequate definition the concept of *probable cost* (chapter 18).

Before we can discuss *limits* and *controls,* though, we need to visit what the architect needs to do to help the owner write the rest of his book.

The Architect's Team

The architect has signed his contract with the owner and has come away with his first reactions to the owner's Project Program. Now he needs to put a team together to produce the documents to match the Program. He assembles the consultants for his design team to participate in brainstorming sessions to refine those ideas, requirements and concepts.

The architect has already translated the owner's goals into project concepts. Next, he has to retranslate those concepts into scopes of work and directions for the other members of his design team. He needs to allocate or negotiate fees. He or she has to make specific assignments and to develop a production schedule. He has to set up and *manage* his design team.

Once he's *translated* Programming requirements into scopes, the architect must become a *manager.* From this time onwards, the architect needs to actively maintain a balance between translation, interpretation, evaluation and management, through the completion of the construction and bid documents.

Once he's translated Programming requirements into scopes, the architect needs to actively maintain a balance between translation, interpretation, evaluation and management, through the completion of the construction and bid documents.

83

There are standards of professional performance the architect must meet in all of these tasks, whether or not they're spelled out in his contract. They are the standards of his profession to which he committed when he undertook to practice architecture.

Experience has demonstrated that the causes of 80% or more of the liabilities to which an owner can be exposed through a construction claim are in those construction documents.

Most important among these standards is that the architect has assumed the duty to produce what the owner will be warranting to the contractor: a set of buildable, accurate and complete construction documents.

Subject to the terms of his specific agreement with the owner, he commits to deliver them within an agreed timetable. He warrants that they will be buildable within the budget to which both he and the owner have agreed.

Experience has demonstrated that the causes of 80% or more of the liabilities to which an owner can be exposed through a construction claim are in those construction documents. It's fair, then, for an owner to consider whether the architect's capabilities are up to the tasks required to produce those documents. And yet, few owners invest the due diligence to determine whether their architects are. Few architects, for instance, are measured for their *management* skills.

The Translator/Interpreter/Team Manager

The architect, whatever his or her creative talents, has four broad areas of responsibility:

- To faithfully translate the owner's Program into construction documents,

- To produce those documents to the standards the owner will warrant to the contractors: that they are buildable, accurate, unambiguous and complete,

- To manage the production of those documents to meet the owner's timetable and budgetary limits,

- To interpret those documents—including their administrative specifications—clearly and comprehensively, first to his design team, and then, with responsiveness and timeliness, to the contractors.

How well the architect performs these tasks will determine whether the measures of *success* we defined in chapter 1 will be met.

The Design Team

The disciplines required to design a project vary over a tremendous range, defined by the complexity and uses to which the project will be put.

A house will require 6 - 12 disciplines: soils testing, civil engineering (site and utilities), structural, mechanical (heating, plumbing, air-conditioning), electrical, lighting, roofing, landscaping, energy analysis, plus investigations into other specialties and codes that the architect may handle himself or delegate to his consultants (safety, finishes, hardware, acoustics, flooring, cabinetry, etc.).

As the complexity of a project increases, so does the need for additional consultants. Special studies and designs may be required for vertical and horizontal transportation systems, production layouts, security, communications, public safety, environmental controls, acoustical enhancement and attenuation, specialized surfaces and finishes, specialized equipment and a myriad of concerns unique to respective projects. The very diversity of projects has required many architectural firms to specialize, while others have developed internal departments for some of these specialties.

It is important for the owner to know how the architect will assign, distribute and structure the relationships of all of the disciplines required on his project and how he will measure their performances.

There may be soils reports to be developed and surveys to run so that their results can be interpreted and incorporated into the project. The project may require extensive test borings (There has seldom been a project that drilled too many test holes.)

On renovation work, extensive destructive testing and exploratory work may be required—no one ever does enough.

Civil engineers, specialty consultants, energy consultants, specifications writers and cost estimators may have to be added to the design team. Applicable codes and regulations control each specialty. Where specialists do not cover an area, the architect will have to research information from vendors and manufacturers to incorporate it into the documents.

The calculations these consultants make—structural, mechanical, acoustical, environmental, budgetary—need to be reviewed by the architect for their appropriateness and consistency with the project criteria, and the consultants must be held accountable for the contributions they make to the project's compliance with the Program.

The architect needs to define how the consultants' scopes relate to each other. He needs to define and maintain the management logistics for how the work of all these disciplines comes together into the documents—what is known as *document coordination*. He needs to schedule their production to meet the owner's timetable, and he needs to confirm that the total assemblage will remain within the project budget.

In the aggregate, the duties of the architect are a major undertaking on even a single residential project. Especially so in terms of the technical, environmental and governmental tunes to which every construction project must dance today.

The weight of these tasks, however, does not relieve the architect of his responsibility to perform all of them professionally. *He is committing the owner to guarantee the documents he produces.*

In becoming an architect, he has assumed a duty to perform these tasks within a professionally defined standard of care. He cannot—despite the wriggling and breast beating heard today from architects, their attorneys and their E&O carriers—escape that duty *as long as the owner meets his side of the bargain.*

That is why it becomes so important for the owner to understand his own duties through each phase of design.

The double review of these phases that we will follow—in this chapter and the next—is intended to serve two purposes:

- To provide the owner with the essential criteria with which to select an architect,

- To provide the owner with the perspectives of his own duties in the development of the construction documents.

Developing Your Project Plan

The definition we've already established for your Project Plan is that it will define *how* you intend to manage your project. It's essential if you're going to build it twice that you have this plan in place as the source for developing your requests for qualifications from architects, and to prepare both their scopes of work and their agreements for services.

We'll cover the development of your Plan, and the management tools you have available to put into it, in Part 9. What follows here and in chapter 9 assumes that you have already put your Plan and your Preliminary Program together and are implementing it throughout the architect's development of the following design phases.

When we review the standard AIA agreements and their mirror images in chapter 18, it will be evident that they contain neither adequate definitions, scopes of work, nor ways to measure the content of the products the architect will produce. They contain no management tools with which to measure or document the design team's performance, and no realistic bases with which to measure payments in terms of design production. Those are the subjects of the systems and procedures we'll discuss in Parts 8 and 9 of this book.

Programming

The standard AIA design agreement requires the owner to deliver a Program to the architect. This is what we have referred to as the owner's Preliminary Program—an effort of the owner that must parallel his development of the Project Plan.

A competent and responsive architect will employ translational psychology to review this with the owner. The architect needs to extract and identify what the owner *really needs,* as distinct from what he thinks he *wants.*

The owner needs to be particularly attentive during this extraction period. He needs to assure himself that he is clearly conveying his requirements. And he needs to be assured that he is not buying into more of the architect's *muse* than he wants, or can afford. The notebook initially recommended in chapter 3 will be particularly useful at this time.

Developing a comprehensive and competent Project Program is a major effort. A program for a single private home may be only a few pages, together with sketches and photos. Programs for complex projects can be inches thick, or several volumes, incorporating preliminary designs, sketches, calculations and layouts, production requirements, circulation patterns, environmental studies and much more, carefully organized by areas and disciplines. Their development can take days, weeks or months, and can require the services of dozens of consultants.

Computer Database

If you're handy with a computer and can master a computerized database, your notebook can be developed as a useful system of checklists you can use throughout the entire project.

8

Whether brief, or complex, a complete program requires a disciplined effort to organize, in as much detail as possible, what will be required in the completed project. The interaction of systems, surfaces, materials (equipment, lighting, communications, etc.), and their relation to usage, traffic patterns, energy levels, access and egress, life safety, production programs and human occupancy can all be significant considerations. A Program may contain detailed descriptions and explanations of requirements, specifications, calculations, budgets, schedules or any other criteria that are essential for a complete definition of the final project.

Maintenance and Operating Costs

Many projects have come a-croppers because considerations of maintenance and operating costs have not been incorporated into the final Program.

Clear definitions of these considerations need to be incorporated into the finalized Program before the owner authorizes the architect to proceed with Schematics.

Owners who have developed comprehensive project criteria ahead of hiring a design team have come to recognize how much this can contribute to a project's success.

Other organizations, notably public agencies, have demonstrated that democracy does not contribute to effective design. In some instances, decisions by committee have led to so much internal maneuvering that the entire process bogs down into unintelligible directions that could not be interpreted into competent design.

When an existing structure is to be renovated or added to, the owner has a duty to provide accurate, complete information of existing conditions to the designers.

GMP: Its Relationship to Program (See also page 76.)

It is particularly important, if the owner is to proceed with a Guaranteed Maximum Price (GMP) construction contract, to develop a comprehensive Program that incorporates standards of quality and that estimates reasonable quantities of significant materials. The GMP is often developed on the basis of incomplete drawings that do not yet show the final details for such items as equipment, reinforcement, hardware, lighting, distribution services, and so forth.

More than one project has had its budget blown away after the GMP was set because the contractor *discovered that "the final design incorporated more reinforcing steel"—or more something—than he "provided for" in his GMP.*

This is a particular kind of hornswoggle that can be avoided by incorporating definitive descriptions and limits in the outline specifications.

Committee decisions frequently result in complete surrender to the architect's *muse* and *chaos* without competent means to confirm design responsiveness. The numbers of projects that are bid with those liabilities still hiding in their construction documents are testimony to the inappropriateness of the process.

What's being suggested here is that:

> *The owner who wants to assure a successful project has to take charge early—through the development of a Project Plan that clearly defines the decision-making process for the life of the project.*

Democratic decision-making works well in some business environments for which the team assigned to the work has relative stability.

Construction is not a democratic environment. It requires a clearly charted course and a strong hand at the tiller before the ship leaves port.

"Excessive Rebar"

The hospital design had reached the completion of Design Development. The contractor—who had "value engineering" responsibility during design—provided the owner with his GMP: $8.5 million.

Two months later, the construction documents were completed. The contractor provided the owner with a revised GMP of $10.5 million. Both the owner and the architect were stunned.

The contractor's explanation was that, among other items, the "reinforcing steel was excessive."

In fact, the rebar was conventional. As was the rest of the design.

But the owner had left itself open, and the architect had undone him by not keeping the specifications concurrent with the drawings, and by not providing a program that, in the absence of details in the drawings, defined in at least narrative form the quality, standards and criteria to be incorporated into the project.

In a protracted negotiation that cost the project a delayed start of three months, compromises were effected through the assistance of an independent consultant. Some design was reworked; some of the contractor's asserted costs were reduced, and the final GMP came down to $9.1 million.

8

The Architect's Negotiations to Put the Design Team Together

Once the owner has resolved an agreement with the architect, the architect has to negotiate subconsultant agreements for the design disciplines required by the project. He puts his design team together in pretty much the same way the general contractor assembles its subcontractors.

It should be no surprise to the owner who has haggled a fee with his architect that the architect will do the same with his subconsultants. It's not usually the subject of polite conversation that the architect will negotiate sharply, and may even "bid-shop" the fees he's prepared to pay his subconsultants.

If, for instance, the construction costs are projected to be $2.6 million, and the architect's fee is set at 8% on this base, there will be $208,000 to be distributed among a dozen or more consultants, including a specifications writer and a budget-cost estimator. These fees might break down as shown in the hypothetical example in the box.

One Possible Allocation of Design Fees:

Construction Budget: $2,600,000

Structural Consultant	1.02%	$26,520
Electrical Consultant	1.02%	$26,520
Mechanical/Plumbing Consult.	1.24%	$32,240
Roofing, Misc. Consultants	0.68%	$17,680
Others	1.36%	$35,360
Cost Estimator	0.48%	$12,480
Subtotal	5.80%	$150,800
Architect	2.20%	$57,200
Total	**8.00%**	**$208,000**

Using this as an example (neither typical nor necessarily realistic for every project, but only as an example of what the owner might consider), about $57,200 would be available for the architect.

Let's say that the average hourly rate billable—for everyone from draftspersons to principals—against this fee is $53 per hour, to cover salaries, non-reimbursable costs, overhead and profit. That means the architect can expend about 1,080 labor hours on the owner's project before eating into his overhead and profit.

If the architect assigns three of his own staff to work full time on this project, the 1,080 hours available will allow about 45 crew-days to prepare the drawings and specifications, meet with the client, confirm all applicable regulations and codes, coordinate and check all subconsultant drawings and specifications, develop at least three cost estimates and prepare the bid documents through all four phases of programming, schematics, design development and construction documents.

The architect may or may not consider that adequate. If he can't negotiate a higher fee, he has two options: to negotiate down the fees he pays to his other consultants; or to short-shrift his efforts to check, coordinate and confirm the accuracy of all of these integrated documents. He may do both.

To the extent that his subconsultants have been negotiated out of what they considered as reasonable fees, their attentions can be easily diverted to other projects.

They may invest minimal effort in cross-checking and coordinating their work with the architect's or structural engineer's controlling dimensions.

Considering all the potential variables, the calculations in the box are provided only as a hypothetical reference to suggest how the design consultants might evaluate the fees the owner's prepared to pay. An exercise along these lines in relation to your project budget and fee schedule may suggest the representation you can expect from your architect and from the design team. A similar exercise may suggest that you need to define an open-book accounting with your design consultants as a basis for measuring performances and payments.

8

DESIGN

Design Production and Payment Schedules

The standard AIA agreement commits the architect to provide a design schedule for the Schematics, Design Development and Construction Documents phases. It commits a "probable" or "estimated" construction-cost estimate at the end of each phase. The typical agreement provides for payment to the architect at the 50%, 75%, and 90% completion of documents for each of these phases.

The standard agreement does not have any definitions of what these percentages mean, or what will be produced in each phase. There is no description of the level of detail to which the "estimated costs" statement will be developed. No standards are expressed with which to measure the accuracy, coordination or buildability of the documents. There are no explicit standards against which the design team's professional performances can be measured in the agreement.

The only standard available to the owner is the one the law has established by legal precedent: *what another professional would reasonably do.*

An owner needs only to visit the territories of the strangers in this book to wonder what that is.

The AIA agreements leave the owner swimming pretty much alone when it comes to what he can expect from the architect. Since few owners either read construction drawings, or understand how to apply the specifications, drowning in a sea of change orders after the CDs have been issued can be a distinct possibility for the owner and his project.

Schematics

This is the phase in which the owner will see in plan view, in elevations and in sketches or colored renderings the generalized schemes and layout of the project. The owner needs to be an active participant in this phase. The drawings will contain only generalized dimensions, but it's important for the owner to check these and to be prepared with some challenging questions. He needs to consider the overall layout and appearance of the planned design against his practical and visceral perceptions—possibly with the assistance of his guide, who should be able to conceptualize better than the owner what will have to be done beneath the finished surfaces to make the project work.

Obvious, but often overlooked, considerations such as wind direction, seasonal temperatures or height of the sun in different seasons can be important concerns affecting energy costs and occupancy. They will be set once the Schematics are approved.

Once the Schematic drawings and outline specifications are completed, the major elements of the project will become pretty much irreversible without at least some extra effort and time.

The architect should be required to produce an adequately detailed outline specification for the major components during this phase. This, again, is a standard that's often honored by omission, since the design consultants will defer the development of specifications as long as possible.

This is also the first time the architect will provide a "probable cost" estimate for the overall project. He will also expect to be paid on the basis of those inadequately defined percentages noted previously.

Design Development

With acceptance of the Schematics and authorization to proceed with Design Development, the architect will start to produce "background drawings" for other members of the design team. These are drawings in which both exterior and interior dimensions start to be nailed down. With these the other consultants can refine their building loads, relationships between architectural elements, structure and distribution systems. Spaces will be defined, and locations for equipment and other elements will be incorporated into the drawings.

Unless the owner requires otherwise, project specifications will typically drag behind. It's not unlikely that when the Design Development drawings are asserted to be 75%, the architecturals will actually be closer to 65%, the structural 55%, the MP&E (mechanical, plumbing and electrical) may be 35% - 40%, and the accompanying specifications will be 15% - 20%, or nonexistent.

There's cause to wonder how reliable the architect's "probable cost" will be when he asserts that Design Development documents are 100% complete. (DD documents that come up to the percentages claimed for them have seldom been seen.)

8

Coordination of the work of the respective disciplines during Design Development is a major consideration. Spaces have been defined. It's critical that everything still not shown on the drawings will fit into these spaces. The only way to accomplish this is to have the specifications, especially for large systems (ductwork, mechanical equipment, switchgear, etc.) far enough advanced that it can be confirmed that they fit—and even that they can be gotten through the doorways!

Once the products and the "probable costs" defined at the end of this phase are accepted by the owner, dimensions will be set and the intensive detailing of the "working drawings" will start.

Construction Documents

The design consultants now start to produce the details with which the contractors will be required to work.

Design Paste-Ups—From Three Projects Back

As JBA continued its pre-construction constructability (B/BA) reviews of the design drawings, curious shadows started to appear around certain details. It would not have been so objectionable if they had been coordinated, but the details of window heads would show up on one drawing, the side frames on another, and the sills on a third and separate sheet.

As they accumulated in successive reviews, it became even worse: the details of one part of an assembly was shown in scale 3/8" = 1 ft., another shown as 1/2" = 1 ft., and a third element as 1/4" = 1 ft. The architect had been borrowing details from as far back as three generations of similar projects.

Fortunately JBA had incorporated into the architect's contract a requirement for consistent coordination of the drawings. Drawings were corrected and coordinated before they went out to bid at no cost to the owner.

In subsequent contracts, JBA refined its requirements to incorporate a standard that *borrowed* or referenced details would only be acceptable if they were incorporated into supplemental detail books as a matter of standard practice, in which case the owner would pay only a pre-defined charge for the use of these details.

Pages and pages of details will be developed (or sometimes borrowed) from details taken from standard catalogs, equipment and materials tables, templates, information from vendors, *Sweet's Catalog* (volumes, standard in the industry, which compile manufactured fixtures, equipment and materials available to the industry).

Concurrently, the specifications will be completed and/or refined. If the owner's not watching closely (by way of the B/BA we will expand upon later), this (particularly in this day of desktop computers) is likely to be downloaded from a canned specification by each of the subconsultants, checked cursorily for general compatibility with the drawings, and incorporated into the CDs.

Here again, scrutiny by some experienced, objective eyeballs is essential.

It's entirely possible that the sets of hands that drew the drawings (or developed them on CAD—computer-assisted-drafting—systems) are not adequately cross-pollinated with the hands that produce the specifications.

After 40 years in the industry, I stopped counting how many doors, rooms, overhead clearances and mechanical chases weren't sized for the equipment they had to contain. As obvious as this oversight is in hindsight, it's almost guaranteed the owner will pay for its discovery during construction.

Also beyond counting are the number of times construction documents contained Rev. 5 of the architectural drawings, while the last background drawings the structural engineer had received were Rev. 4, and the systems consultants had referenced Rev. 3. The results are drawings replete with the physical conflicts and "failures to coordinate" we heard about in court (chapter 6).

And then there's the front part of the specifications: the "boilerplate" and "special requirements." The administrative part of the specifications that will instruct the contractor *how* he needs to perform to be responsive to the owner's administrative requirements.

By the time you have developed an effective Project Plan, these administrative specifications should be an evolved product of the relationship you have established with your guide and your attorney. They should be in process before you retain your architect.

8

The alternative is that they will be the AIA A201 general requirements for administration of the construction contract (or its equivalent). By then, the fat's right back in the fires that await in the Black Hole.

We will be discussing the *boilerplate* and its many ramifications in Parts 7, 8 and 9.

Bidding and Bid Documents and Conferences

Construction starts once the construction documents go out to the printer. Anything that happens after that (we've already seen how contractors view addenda) often ends up as a change order or worse, a claim.

The bid documents are produced during the Construction Documents phase. They're an extension of the procedures and contract requirements the owner should have set up in the Plan. It's in the bid documents that the bidders are instructed how to submit their bids, what documents must accompany them, and what information the owner requires in order to make an informed selection of a contractor.

The type of contract may require unit-cost breakdowns, breakdown of subcontractor bids, establish special bonding requirements, stipulate terms for award of contract, or any of dozens of other special considerations relating to the bids.

They will define the time for return of bids, and requirements for pre-bid conferences, if any.

The first evidence that there may have been some chaos among the design consultants during design shows up if the architect finds it necessary to issue an addendum that makes design changes. Sometimes it's not a fault of the architect's design. The owner may be the culprit. In any case, the red flag's been run up the pole for all to see.

Continuing evidence of ineptitude can surface during the pre-bid conference. Surprise or hesitancy at questions raised by the bidders starts to telegraph the smell of blood.

Inept handling of the bidding phase, demonstrated by delays in responses to questions raised by any of the bidders, inadequacy of answers, responses without written confirmation to all bidders, are all signals for potential disputes.

Once a successful bidder has been identified, delays in award of the contract, failure to provide timely access to the worksite, or equivocal reactions to proposals for changes or substitutions can all send signals of potentially inept contract administration to the contractor.

By applying to the bidding procedures the same anticipatory evaluation that he has put into the development of his Project Plan the owner can avoid these liabilities.

CONSTRUCTION

He may not tell you so, but the architect's participation during construction—even if he has no responsibility for contract administration—is central and critical to construction and to the contractor's own schedule of performance.

8

Even when—as *BIT* recommends—the responsibility for preparing change orders, negotiating costs and payments to the contractor, evaluating schedules and determining the resolution of disputes are removed from the architect's scope of work, he retains a critical duty.

He retains the duty to assure design compliance. This means that he reviews and "accepts" submittals on behalf of the owner. He responds to design questions raised in RFIs. He must be present at regular, and sometimes special, meetings that involve design decisions. He prepares the design components of change orders (the resolution of time and money have been retained by the owner). The architect has continuing responsibility (whatever his "periodic observations" amount to) to see that what's built is what he put in the documents.

The timeliness, the adequacy and the accuracy with which the architect accomplishes these duties can have major implications for the *use of time* on the project. His failures in these areas have often been the hidden source of time-related disputes. Hidden because, without adequate documentation of the architect's performance in the owner's files, *fault* becomes a contest between the architect and contractor—a contest into which the owner is inevitably drawn because the construction contract is between the owner and the contractor.

In the standard AIA General Conditions for construction (AIA A201) you will find, if anything, a *lesser degree* of specificity regarding the architect's standards of performance during construction, than in the terms of the AIA B141 regarding his performance during design.

The architect and his products are at the center of the construction universe. But you wouldn't know it reading through either those standard AIA General Conditions or the other documents that purportedly define his duties.

The architect's contributions during construction to the success or failure of a construction project are so pervasive that we can't cover them all here. It will be productive, though, as you develop your Plan, to reread the processes that require timely and responsive interchange of information between the architect and the contractor, as described under an Early Warning System in chapter 5.

Submittals

When the contractor delivers shop drawings, the architect must distribute them to the appropriate design consultants, log their processing, schedule and coordinate their return, and ensure that the information returned to the contractor is timely and accurate. If he rejects them because the design was faulty to begin with, or because the specifications are inadequate to measure their design compliance, it could start a train of events that can go all the way to a construction claim.

The contractor assumes the duty to produce field drawings, shop drawings and submittals that demonstrate that his work will comply with the intent of the documents. They must, as examples, define how the plumbing mechanical and electrical systems will be installed in order to comply with the design. They need to define the specific equipment, structural connections, types of materials and other components the contractor intends to install.

The consultants responsible for the design of specific disciplines may accept them as submitted, may mark them up with clarifications, may require that they be modified and re-submitted, or may reject them out of hand. The architect is charged with coordinating these responses into a cohesive package and return it, timely, to the contractor.

If any element of this process exposes design defects, is delayed, identifies problems, or is incompetently handled by either the contractor or the consultants, it can be the source of some of the mostly costly impacts to a project's budget and schedule.

If the cycling of submittals is not timely, the construction can be delayed. If the contractor has discovered a non-buildable element in the drawings, it can result in demonstrating faulty design. If the architect or consultants stonewall or deny responsibility, there can be a dispute that drags on for weeks or months.

If the submittal is faulty and the contractor has to re-submit a detail (out to the subcontractor, back to the contractor, back to the architect, to the consultant, then back again all the way to the subcontractor), this can also create delays, arguments, RFIs, changes and a change order. Even if the contractor's original submittal was faulty, it can take months to sort out the ultimate

8

responsibility, *unless the owner has received the documentation that records these transactions and has someone who can assist with objective evaluation of the facts.*

The ramifications can sometimes be mind-boggling, and we can't cover them all here. Some of the war stories will hint at the possibilities.

Requests for Information (RFIs)

The next test for the architect may come in the form of a Request for Information (RFI) from the contractor. The contractor(s) may periodically request clarification of something that's not clear in the construction documents, or something that *is* clear, but about which the contractor has decided to make a fuss (possibly to cover his own mistakes). It may result from information missing in the documents that the contractor asserts he needs in order to meet the intent of the construction documents, or from an ambiguity, or from a conflict (two or more things designed into the same physical space), or from an outright error (dimensions that don't work).

The response to an RFI may have to be carried through all the same steps as a submittal, or possibly all the way back to a vendor (to check data of a manufactured item).

Even if the answer to the RFI is benign (results in information only, without the need for a change order), the process can consume valuable time. If the work that requires the information is "critical" or "controlling work"—that is, delaying it will delay the overall construction progress—the architect's timely response can be important to keeping costs within the budget. If the architect

---

FPO—Field Proceed Order

Time considerations often make it essential that the contractor be able to proceed with work in less time than a change order can be approved and executed. The FPO is a reasonable way to get the work going. In a sense it is an interim change order and, as such, should be comprehensive enough to describe the scope of the work and the limits of the costs that are being authorized. It must represent a clear understanding of the parties who will eventually execute a change order for that work. Anything less is dangerous—and anything less is all too typical on many projects.

delays, it can be the first place the relationship between the architect and contractor starts to tear at the edges.

The architect's response will be significant in terms of the continuing relationship between the architect and contractor. This will be true—and can reflect on everyone else—regardless of who's managing the project: architect, CM, contractor or owner.

If the answer to the RFI does require changes, the processing of the change order may take some time, which raises the challenge of how best to keep the work flowing in the field (supplying the answers), even while the change order remains to be formalized, negotiated and authorized. This is where a Field Proceed Order (FPO) might be called for.

The contractor may just be playing games. Sometimes he has a legitimate question, and the architect realizes he's goofed. Either way, delay only compounds the problem and compromises the owner's budget.

An RFI can lead to a change order, but the worst thing that can happen is that response is delayed until the result is the assertion of a delay claim. That adds to the potential cost of a condition that might have been unbuildable to begin with, but should at least have been resolved expeditiously.

The owner needs to be represented by someone who can independently and non-defensively ensure that timely, objective responses are provided to these questions. Objectivity is essential to defuse potentially budget-destructive situations, even if they might add extra direct costs to the project.

It's the consequential costs that'll getcha. All of which brings the owner back to consideration of the leverage and the indemnification (E&O coverage) from the architect the owner has incorporated into his agreements (see chapter 12).

Change Proposals
and Change Orders (C/Os)

The terms of almost all construction contracts require that a written change order be executed by both parties before the contractor can proceed with the change. Otherwise the contractor may be risking that he will not be paid for the extra work.

The requirement that a change order be executed before the changed work can begin is as often disregarded as observed. The time it takes for some owners (especially public agencies) to authorize changes is not consistent with keeping the Work on schedule (the contractor's overall scope of work is spelled with a capital "W" in the contract). This written requirement is often circumvented with an intermediate *proceed order,* or *field proceed order* (FPO). The critical consideration here is that this intermediate order adequately defines and limits the scope of the change, and puts a fence around the amount of money that will be expended.

The contractor may take the risk, as long as he has established a reliable relationship with whoever is administering his contract, based on an oral assurance that the paperwork will follow. Even an oral directive, however, must be documented to the owner by the contract administrator.

More often than it should happen, payment for the extra work the contractor presumes he will be paid for is repudiated by the architect, or by the owner, because they have somehow decided (or had intended all along) that the burden should be put back onto the contractor. This set of circumstances can get very messy before it's cured.

It's always useful for an owner to remember, however, that the contractor has more ways to mount a claim, or hide his mistakes, than the owner, the architect and the CM, together, will be able to imagine.

That's why *BIT* recommends that honesty and candor are among an owner's best resources. Exposure of his budget to a protracted claim is not the ideal way for an owner to realize what can happen to his project and his budget if a contractor decides he's not being treated fairly.

ASIs

An ASI (Architect's Supplemental Instruction) can be a benign and expeditious way to handle a question that comes up during construction—or it can be the insidious beginning of a dispute. The appropriate function of an ASI is to provide instructions to the contractor regarding how to perform a certain part of the Work from which there will realistically be no impact on time or to the contract costs.

If that judgment is made objectively by the architect, and accepted as such by the contractor, that should be the end of it.

On all too many projects, however, the architect tries to pass off an ASI to make a design correction that *will* have time and/or cost implications. Only the best of established relationships between the architect and contractor will treat this latter condition benignly, and only then if there is some potential for "horsetrading" later to balance accounts.

In all cases, the owner needs to be apprised *as it is being processed* of everything he needs to know to make his own independent judgment of what a specific ASI (or RFI) might do to his project.

EWOs

Many contractors will "open the book" by tracking Extra Work Orders (EWO) for each event or condition they think might cost them money.

Some less-than-scrupulous contractors will record even a cursory remark—or a sneeze—by one of the design consultants at a progress meeting as a potential EWO. At the end of the project, the ubiquitous *claim consultant* will incorporate into the contractor's claim history and into his *"As-built CPM analysis"* that the contractor *"experienced a total of 327 EWOs on the project!"*

It's in the owner's interest to have the contractor provide his list of EWOs on a regular basis at progress meetings so they can be talked out, resolved, deleted, if possible, or, in any event, anticipated.

The Constructive Change Order

When conditions in the documents or on the project require that extra work be done, but the owner refuses to pay for it, the contractor is challenged to establish that he was working under a "constructive change order."

In these circumstances, arguments regarding the contractor's proposed costs may lead to protracted delays—delays that may or may not be justified. The arguments about delays, costs, and responsibilities that can surround a constructive change are too numerous to list. The only way an owner can sort them all out is to have access to all of the documents, concurrently with the events that generate them.

Mitigation

There is a principle well recognized in construction law that says that a contractor cannot take advantage of a problem on a construction project to run up the costs or gratuitously expand either the problem or delay the completion of the Work.

8

The contractor, nonetheless, is under a reciprocal duty to "mitigate" the costs of extra work for which he is likely to claim later that he has operated under the conditions of a "constructive change order." He can be in breach of contract if he unnecessarily and unjustifiably expands the time or costs beyond what's required. (See page 50.)

Heading Them Off at the Pass

As all of these circumstances demonstrate, issues revolving around the interactions of the contractor and the architect can become very complex, very damaging to a project, and very costly.

It's fair to say that there never was a project that escaped these problems altogether.

The owner needs to stay on top of them so that a scratch doesn't become infected, and so that the wound doesn't lead to amputation.

In the order of their declining effectiveness, the ways for the owner to do that are:

- By providing a clean set of construction documents.

- By encouraging and facilitating open, candid, honest working relationships among all the parties right from the outset of design, and during construction. Cooperative people can solve many problems with minimal liabilities.

- By delivery to the owner of competent, comprehensive and objective documentation of the events and interactions on the project, along with effective use of those documents.

The owner's tool for assuring these goals during design is the B/BA.

During construction the owner needs to receive all transactional documents, from the first submittal, RFI, ASI or request for a change, so that the issues of responsibility for changes, and for the timeliness of responses, can be acted upon before they interfere with the schedule or the budget.

Even if the owner must ultimately pay the price of having issued documents that require costly change orders, it is in his interest to know early in the process where matters lie and to solve those problems quickly. The alternative is to have them disappear into the design consultants' or contractors' back offices, only to surface when they have already, and unnecessarily, delayed the overall project.

This is one of the critical areas for which the owner needs to have the experienced eye of an independent guide assisting him with objective evaluations of responsibility.

The Construction Schedule

Meanwhile, the contractor will have been required to prepare a schedule that shows how the contractor plans to prosecute the Work in his contract.

Someone has to evaluate the reasonableness of this schedule on behalf of the owner. Whether or not that someone is the architect, the architect must understand and participate in its evaluation because his review and acceptance of the contractor's scheduling of submittals will be pivotal to the performance of this schedule. His reviews will need to be performed within a time frame that allows the contractor to work without interruption, in particular on "critical" work.

The aggregate of all the perceptions about, and competence (or incompetence) with, scheduling has become a long, sad tale for those of us who believe that the management of time is of critical importance to the successful direction of any project (construction or otherwise).

It is especially critical for a project on which a couple hundred contractors and vendors rely on the work of a dozen or more consultants.

The issue of *time management* is so important that I've committed all of Part 6 of this book to it.

It's sufficient to note for now that there is a sad lack within the design professions of either a commitment to, or understanding of, the disciplines associated with effective project scheduling.

8

Inspection

The architect has the responsibility throughout construction to review the Work for its compliance with the construction documents, and with the "standards of the industry." Some public agencies, however, assume this responsibility after delivery of the CDs by the design consultants (a sometimes questionable policy).

In the area of inspection, the architect may be assisted by an independent inspector or inspectors whose duty it is to ensure and record that the Work installed meets the standards defined in the construction documents.

The inspector who may occasionally come to the project from the offices of a public agency is not specifically charged with protecting the owner's interests. He does not perform the inspection of the Work the owner needs.

The owner requires an inspector (or several) retained specifically to protect those concerns.

> On some projects, notably schools and health facilities this inspector serves two masters—the owner and the state—or even three, if he has a long-term relationship with the architect. The potential conflicts of interest in these relationships have frequently led to disputes with contractors, who are caught in-between.

The competence and integrity of the owner's inspector can be critical to maintaining the integrity of the construction documents, since much of the work of construction will be buried, hidden, covered up. The inspector's view of what and how it was installed may be the only record of whether it complied with the contract.

The concerned owner should have a hand in the selection of the inspector. He should receive copies of the information collected and transmitted by the inspector to the architect.

If the inspector is competent and committed to the owner's interests (as distinct from any other party on the project) he can be an important source of information. The owner can take advantage of the inspector's daily presence on the job to request objective reporting of the overall performance of the Work by both the architect and the contractor, along with his reporting of other areas of concern.

Saved During Design: $175,000 and 2-1/2 months

The Design Development drawings had just been completed when the owner authorized the Owner's Representative (OR), who was already working on the administrative specifications, to undertake a constructability review of the documents for the $20 million hospital addition. The project was five months away from going out to bid.

The OR started development of a pre-bid scheduling analysis and early on determined that the emergency generator, which would go in the basement, could not be delivered until the structure (concrete columns *and* concrete walls) would likely reach the 3rd or 4th level of an eight-story tower.

The OR proposed that the basement walls incorporate a knock-out panel; and that the bid documents include provision for a trench into which the generator could be dropped and skidded into the basement.

Once construction was underway, the contractor acknowledged that he would have change-ordered this arrangement at a cost of about $125,000, but had allowed only $20,000 in his competitive bid.

In the pre-bid analysis, the OR also estimated that by changing the concrete specs from 3,500 psi concrete to 5,000 psi concrete, about 2-1/2 months could be saved in the cure-strip-and-reforming duration of the concrete tower. At an estimated cost of $40,000, the owner concurred that this would provide a benefit to the project.

The contractor later acknowledged that it would have tried to collect about $85,000 for this change during construction, but had put an allowance of only $15,000 into its bid.

One other subtlety about inspectors: the inspector can be a lonely person. He is alone on the job for much of the time. His only continuous contact is with the contractor. This relationship has more than once interfered with the objective reporting the owner deserves and believes he is receiving. An owner needs to recognize this nuance of the inspector's role on a project.

The matter of the *inspection* and the *inspector* is a subject unto itself. He or she can be an important asset to a project—or a destructive force. It is a subject that an owner should discuss extensively with his guide.

8

Post-Construction

In addition to the standard forms—the Notice of Completion, final payment, lien releases, etc.—the owner needs two sets of other documents from his design consultants once the job is complete.

One of them is a *Close-Out Report*. This should incorporate a comprehensive recapitulation, moderated by the owner's objective representative, and with contributions from the architect and his consultants, of the changes, relationships, reports, status of disputes and any other salient information generated during construction.

If the comprehensive documents recommended in *BIT* have been maintained, this can be accomplished with overviews and annotations of the documents already delivered.

The other essential documents the owner must define and require are the As-Built Documents described at length in chapter 7.

Fresh Eyes

Before computer programs allowed you to walk or fly through your future project, most owners who wanted to sense how rooms would look and how spaces flowed through their building had to wait until the sheetrock was up.

Even with these aids, though, it usually takes a practiced eye to see the physical relationships, elements and construction details that may or may not work out.

The most valuable resource an owner might associate into his or her project is a fresh set of eyes that come equipped with the experience, the objectivity and the unblunted perspectives to see what may have been swallowed up by the bustle of design-office production and owner distractions. (The Wakeup Call on page 120 and the story on the previous page are representative.)

9

THE OWNER AND
THE DESIGN CONSULTANTS

Let's Find An Architect

It's useful to start our search with the products we expect the design consultants to produce:

> *When the owner executes the agreement with the contractor he warrants not only that the construction documents are complete, accurate and buildable, but that the contractor will be provided the time and the conditions necessary to perform the Work in those documents without interference. The owner also warrants that timely payment will be made for extra work, as well as base contract work, properly performed.*

One of the rights the contractor acquires with his contract is the right to build the project in any reasonable way, unless that has been specifically restricted in his contract.

Time is the contractor's resource. It costs both the owner and the contractor money to use that resource. The amount of time the contractor has, and the means both he and the owner will use to plan and measure it, are defined for better or worse once the construction documents go to the printer.

Stranger

When you read the description for *stranger* in the glossary, you may have thought it an exaggeration.

But if the performances of the firms you hire will have major implications for your economic well-being for the next 3, 5, 10 or 30 years (taking operations, maintenance and financing costs into consideration), there is excellent reason to apply the same standards of performance to those firms that you would for someone to whom you'd turn over your investment portfolio.

Despite the *gemutlichkeit*[1] he extends to you, the architect you're about to hire is more a stranger to your way of thinking than you realize.

---

[1] *Gemutlichkeit: (noun, Ger.) cordiality; friendliness.*

9

Question an architect you're considering about his or her experience with construction logistics, his understanding of economic realities, his facility with cost controls, his familiarity with production analysis, the scheduling of resources or about the management capabilities we reviewed in chapters 7 and 8. You are likely to learn that you will have to supplement the resources that most architects are capable of bringing to your project.

Friend

The architect has a large interest in transitioning quickly from stranger to "friend." Once he's made it into that insulated area, it's virtually impossible for an owner to come down on the side of tough performance standards, or to insist upon terms and conditions designed to protect the owner's interests.

Nobody enjoys confronting a "friend."

That's why the time to set up those challenging questions, and the means to measure the answers you receive, is before issuing the architect's RFQ (request for qualifications).

How Did You Find Your Architect?

You will have had several months to discuss the possible formats for finding a contractor[2]. You'll have some idea of their capabilities and track records.

You already have ideas about how you will find your contractor. You may have continuing relationships with contractors you've worked with before and you'll invite them onto your select bid list. Or you'll create a list based upon recommendations from the architect and others who know the industry. Or you'll put your project out for open bids.

How will you select your architect?

Did a friend recommend him? Did you see her work in *Architectural Digest?* Did you get his or her name from your dentist? Did you meet someone over the weekend or at a conference who knew someone who knows an architect?

---

[2] *Unless you've decided to go with a design/build contract, which we visit in chapter 19.*

THE OWNER AND THE DESIGN CONSULTANTS

Will you hire your architect to design a multi-million dollar airport expansion because someone at the club knew he'd designed the ticket counter for his favorite airline? (This happened.)

Will you be impressed with his brochure, the pictures of her recent projects, the design awards he's won?

Have you had several architects schlepping around for months, or years, making "friends" with everyone in your district, agency or department? So that it was just a matter of going through a formularized dog-and-pony-show before you selected one? (A classic selection procedure among school districts.)

> One high school budgeted at $40 million won national awards for its design, experienced 176 change orders, cost $54 million to complete, and was occupied a year late.

Once you put the architect on your interview list, did you give him/her 45 minutes to make a pitch, fill in the blanks on a pre-printed form and add up the numbers?

Or did you tear up those canned interview sheets and discuss in agonizing detail the budgets, addenda, design changes, schedules, ASIs and history of problems on his last 6 or 10 projects?

Did you visit, along with a construction-knowledgeable guide to help ask meaningful questions, the contractors and superintendents on his last several projects?

There is one way to establish whether the architect who *looks* best after the first interviews has the competence to deliver a successful project. That one way is to meet with as many as possible of the owners, contractors and subs for whom, and with whom, your potential architect has worked over the past several years.

The information you're looking for is:

- Whether past projects were delivered within their original budgets—and if not, why not. How many change orders? What areas did they cover? (Distinguish between owner-initiated and design-problem initiated.)

- Was the architect intelligent, responsive, accurate and timely in his answers to contractors during construction of those projects?

- If someone else (a "construction manager") managed the project, how did that relationship work out?

9

- Did the owner receive effective and intelligible information with which he could make timely decisions and take effective action?

- When the design consultants were caught in design errors, how did they respond?

The Follow-up Interview

Then, it's time to confront the design team of choice with reality. With your Plan and your independently acquired information in hand, and with the architect's entire design team present, it's time to tell the team how you plan to administer their contract.

If "confront" is a difficult concept for you, it might help to consider some of the events that can occur during construction. Contractors are masters of confrontation. If your potential architect can't stand a little heat in an interview and still be responsive, you're likely not to be well represented later.

If "confrontation" isn't a comfortable concept for your staff or your selection committee, you might need to remind them how it will pale alongside the agonies hiding in a set of bad construction documents, the confrontations they'll be faced with—and all the new strangers they'll meet—if the architect ducks his head and hides out when his "stuff" hits the fan during construction. The confrontations that will arise if your architect or contractors bring along a bunch of sureties and attorneys, or the agonies your board members will experience if they have to go back to the well for more money.

There are, after all, precedents for bringing your entire design team together before executing their contract, and alerting them to how you intend to manage that contract and their design performances. You will be doing those specific things with contractors in the pre-bid and pre-construction conferences.

If quality performance and timeliness are critical to construction, how can the performance of the team that produces the construction documents be any less?

Why Don't More Owners Manage the Design Process?

- Because there's a misconception among owners that *contractors* are the principal source of construction claims.

112

THE OWNER AND THE DESIGN CONSULTANTS

- Because most owners don't realize that 80% or more of the liabilities to which projects are exposed during construction are contained in the construction documents.

- Because of politics, indifference or inertia.

- Because of the subtly intimidating relationship the architect creates. It's part of the mystique of architecture that the architect is a friend, above the motley business of business, and that, as a professional he will objectively represent the owner in all things.

The AIA strives mightily to reinforce this fantasy. There is an article in the AIA A201 General Requirements that says:

in the event of a contract dispute between the owner and contractor the architect will exercise fairness in judging their respective responsibilities.

This absurdly disregards the fact that it's the architect's products that cause most of the disputes to which the owner and contractor become parties.

While the AIA has half-heartedly modified this article in the General Conditions over time, many owners still buy into the mystique of the architect's ostensible objectivity.

The architect, meanwhile, asserts that it was the contractor's obligation to find and correct mistakes in the construction documents at no cost to either the architect or the owner.

This misconception of the architect's objectivity will be quickly dispelled once an owner attempts to recover from an architect's E&O carrier monies paid to a contractor because of design errors.

Architects are contractors. *Their performances should be measured the same as other contractors'.*

Management Skills

A large number of architects lack the management skills even to coordinate effectively their own subconsultants and products, let alone the performances of contractors.

The reality is that many consulting engineers recognize this, although, most of them won't admit it openly because their livelihood depends upon their relationship with the architectural profession.

This recently, however, became a public legal issue. The NSPE (National Society of Professional Engineers) has brought a legal action against the AIA, asserting that the AIA has improperly shoved engineers aside for work which more properly, and in many cases more effectively, should be headed up by an engineering consultant rather than an architect. The NSPE lost one round, but the battle continues.

NSPE Versus the AIA

The National Society of Professional Engineers has brought a legal action against the American Institute of Architects alleging that the AIA unfairly acts to keep engineers from providing services as the prime design firm on work for which the engineers are qualified and provide the principal design work.

In the first round, the AIA apparently prevailed. But the dispute continues under appeal.

Let's Go Through Those Phases Again

We've discussed in general outline how design services progress through the Schematics, Design Development, Construction Documents, Bidding and Construction phases.

Now it's appropriate to emphasize what the owner should expect his architect to produce in each of those phases, and what the owner's reciprocal duties are.

The Owner's Plan for Measurement of Design Production

By the time the owner is prepared to authorize the start of Schematics, the owner, with his Project Plan in hand, should have resolved with the architect a document that will communicate to the owner, and to all the prospective design consultants:

- Identification of all the design firms on the team together with their qualifications, and the names of the lead persons at each firm who will work on the project to completion of its design.

114

- Specific descriptions of the products deliverable at specific intermediate milestones, and at the end of each phase. The most direct way to accomplish this is for the architect to develop the Index of Drawings and the Index to the Specifications early, and identify exactly what drawings and sections will reach what level of completion within each phase.

- Definition of the level of detail to which the "probable cost" estimates will be developed for each phase. The best way to accomplish this is to require a line-item breakdown form that will be completed at appropriate milestones.

- Description of exactly how the architect will develop the projected construction-contract duration.

- Agreement by the architect to a program whereby the architect will deliver a *checkset* of drawings and progress specifications to the owner for his independent review. The owner will conduct his own independent B/BA of each of these checksets.

- Agreement that the owner's review will result in meetings with all the assembled major design consultants to deliver his assessment of the adequacy of the documents and their adherence to the owner's Program at each of these checkset stages.

- A statement committing the architect to the standards of completeness, accuracy and buildability to which the documents will be delivered to contractors, plus compliance with the budget. (Watch out for the architect's E&O carrier on this one.)

- A program that provides for the owner to meet periodically with all of the design team members to review their designs, and to measure their payment in relation to how well their performances meet the owner's Project Program.

Initially, the budget and schedule projections will be preliminary. As development of construction documents progresses, one of the most important things an owner can do is to ensure that both the schedule and the budget are adjusted, cross-checked and refined to reflect the increased detail of the design documents. By the time the project is ready to go to bid, the owner will have built the project on paper, to a level of detail close to what a contractor will prepare with his bid.

Swinging From the Light Fixtures

There is a county detention facility that has, with a 20-year history of efficient operation, demonstrated the value of intense interest on the part of the owner in all aspects of design and potential construction during the planning and design phases.

As planning started, the sheriff established a project office managed by his own officers. As design considerations evolved, staff participated in discussions of options and provided critiques of different means of accomplishing the project's goals—all without fear of censure and with a pretty good prospect that even off-the-wall ideas would be considered.

As construction details were developed, potential elements, such as light fixtures, wall surfaces, locks and furniture, were installed in the office. Staff pummeled them, kicked them, hung from them, until there was assurance all around that they would meet the department's standards.

It is a facility that—entirely apart from its function—is a model of how a project should be planned, programmed and designed.

Transition

The first of the bridges that lead from Programming to construction is Schematic Design.

Schematic Design is the architect's license to visit his *muse*.

That's the muse that has been germinating in the architect's breast since he or she first started sketching. If he is really creative, the *muse* has become more powerful with each project, with each year he has practiced his profession. And that's good.

It's just that the owner doesn't need to underwrite any impractical expressions of the architect's creativity, unless they meet his own purposes and he's willing to pay for them.

The architect's creativity is one of the two specific resources the owner is looking for to enhance his own ideas and goals. Imagination to enhance practicality. The other, of course, is the architect's ability to assemble all the required disciplines into a cohesive whole.

But if the owner has no clear ideas to begin with, the *muse* can take over, at significant cost to the owner's budget. Unbridled expression of an architect's muse in the Schematic phase can result in an irreversible dilution of an owner's long-range control over the destiny of his project.

That's why it's so important for the owner to have established comprehensive administrative procedures for the management of his project through his Project Plan and preliminary Program. The owner needs a system of clear, concise guidelines for what he wants, what he doesn't want, and what it should cost, before sitting down with the architect's muse.

The owner's Program criteria don't need to—in fact should not—put hard walls around the architect's creativity. But they need to be clear about what turns the owner on, and about what he absolutely does not want.

The owner's Plan, on the other hand, needs to state clearly what standards of performance he expects from the architect. It should define how his work will be measured and paid ("75% Design Development" is meaningless). The production schedule the architect delivers should identify the labor hours assigned to perform tasks. The owner can then establish a realistic basis for the payments that will be made for identifiable documents.

Owners have come to believe they can't establish these guidelines before hiring the architect.

The histories of successful projects, though, demonstrate that their success has been in direct proportion to how much homework the owner did before signing service contracts.

The histories of unsuccessful projects have demonstrated that it was lack of clear definitions, or acceptance of canned contracts and administrative procedures, that got them into trouble.

Schematics

Whether the architect has the ability to manage the design to meet these standards—and whether his insurer will stand behind his commitment to those standards—is something every owner must know before executing a design contract. The design-production plan and schedule need to be delivered to the owner before he should even consider authorizing the

9

$27 million project—0.6 of 1% in Design Changes

The organization's design and construction program was administered by architect-project managers. The organization had, as result been using carbon copies of the AIA contracts for many years.

For three years various in-house teams had made multiple studies for a new distribution center, but had made no progress toward establishing a program. Action had to be taken and management decided to outsource for an Owner's Representative who would work alongside the on-staff project manager to assist as "coach." The OR had authority to recommend any changes that would facilitate the early completion of the project.

The OR's first recommendation was to change the contracts for this project. Working with legal counsel, the OR revised the contract documents to set up a system whereby the general contractor would be retained on a C+FF contract when design reached the DD phase. (See chapter 19 for C+FF.)

The OR then arranged with the project manager to set up a series of weekly meetings in which the various department heads would meet to program and coordinate their requirements. The OR introduced a computer database to organize and integrate this data into the Design Program. Long lead items were identified and investigated even before design started.

RFQs integrating this information were issued to 10 select design firms. One was retained on a production-based, fixed-fee basis. As design approached the Design Development phase, a contractor was retained from a select list of eight. The architect participated in the selection.

With the contractor's fee and reimbursable costs defined and as design progressed, the owner, architect and contractor proposed selected subcontract bidders . The project proceeded on a partial fast-track basis with the contractor participating in the design process. A detailed analysis of construction logistics and scheduling proceeded as part of the process. All remaining Work was competitively bid with contractors required to recognize the schedule in their bids.

The project was completed as originally scheduled. Design-required change orders added a total of 0.6 of 1% to the original budget.

"Save Us from Our Architects"

For many years a close friend has been vice president and contracts manager for a successful and reputable general contractor.

In the past few years, his firm has expanded its operations to provide Program Management as well as general contracting to school districts *on a sole-source basis*, incorporating within its scope of work the management of design services. His firm has saved several school districts hundreds of thousands of dollars, typically delivering projects within 3% or less of their original budgets.

The firm initiated the service when a local school district came to them with the plea: *"Save us from our architects."* The district had come to realize both the self-serving character of the design contracts they had been authorizing and the growing incompetence of design consultants generally.

On the basis of the firm's performance, several districts joined to promote legislation designed to allow the districts to identify on a sole-source basis services they have determined will protect their school programs from deficient design and inadequate design contracts.

---

start of Schematics. The standard AIA agreement doesn't provide any criteria for this level of management.

The Schematics phase should produce a set of drawings and outline specifications sufficient for the owner to become comfortable that the ultimate design will meet the Program requirements and that the architect will commit to a budget the owner can live with.

The owner will be required to make continuing and reliable decisions as part of this process. This is where many projects encounter problems, because the owner believes he can hem, haw,

9

The Wakeup Call

He and his wife had been collecting ideas in a scrapbook since they married five years ago. They had bought the lot two years ago. It took them a year to find an architect. Now they had the drawings in hand and were about to discuss a construction loan with the bank.

Since he was an engineer he had developed a pretty detailed cost estimate and a well defined schedule he expected to work out with a contractor.

While it's contractually important to leave scheduling initiatives to the contractor on a publicly bid or larger, complex project, it's not a bad idea on a smaller project such as a residence to work closely with a contractor. Especially one who may not have the organizational skills that larger contractors (should) have to plan and order materials and stay ahead of the power curve.

Even though they were ready to go, they thought it might be helpful to have an independent viewpoint of the drawings. They left them with a friend who regularly undertook B/BA analyses for clients.

It only took a couple of hours to identify several concerns. The problem was: *how to deflate a friend's balloon?* But since the concerns could have long range implications, the friend took the bit in his teeth and pointed out:

- The garage barely accommodated the length of an automobile. But the couple bicycled. And skied. And liked to take picnics in the country. And traveled occasionally. There was no room in the garage for storing the equipment they typically loaded into their car; or for extra suitcases.

- They had two fireplaces, but no ready access or storage for firewood. It was 120 paces (about 180 feet around the house and through the front door) from the nearest location in which firewood could be stored.

- More importantly, the very pleasant tumbling fountain the architect had designed into the patio at the front door was too close to the entry doorway to allow large furniture to be moved in and out. There was no place for guests to stand out of the rain while waiting at the door. And the fountain represented a convenient place for unfriendly people to hide.

- Most important of all: the wife enjoyed family cooking and had specified that she would like a platform kitchen that overlooked the family-eating area and great room. It did, and provided what she wanted. But no one had picked up on the fact that the other side of the kitchen was open to the hall and stairway that led upstairs. An appealing juxtaposition of spaces on paper... a disaster unless they wanted the smells of burnt toast, bacon and roast lamb throughout the entire 2nd floor.

The drawings went back on the shelf while they pondered their options.

defer and procrastinate. He can. He needs to realize, though, that the cost to his project can include extensions to the architect's contract, conflicts for the contractor to discover later and the kinds of costs you can read about on page 188.

Depending upon the owner's resources, it may be time to ask the guide for help, in order to benefit from independent perspectives of what each successive Schematic Design checkset might imply for the future. The more complex and extensive the project is going to be, the more intermediate reviews the owner should plan.

Owners often place inadequate importance on Schematics. If the colored renderings look good, the doors are in the right place, and the square footage meets production requirements, that's as far as they go. But it will soon be too late to make adjustments without additional costs, or at least frustration on both sides of the design process.

Schematic design is an excellent time for the owner to start learning how to read his own design documents in any case (we'll discuss a method for this in chapter 23).

Design Development

It's during Design Development that any facility the owner has developed for reading drawings and specifications will deliver its greatest value.

Design Development is early enough to make changes without incurring significant additional design costs (if any); and it's just about the last effective opportunity to do so. If the owner can develop a facility (or find a guide to assist) for translating flat lines into three-dimensional conceptualizations of the final project, this is the time to do it.

Conflicts between the drawings and the specifications are often a major source of problems during construction. Specifications are often written by someone who has not followed along with the thought processes that have gone into the drawings. This can happen with any of the design disciplines, not just in the architect's offices. Many design offices fail to assign someone whose job it is to check frequently back and forth between the drawings and the progressive specs to ensure that they're complementary, consistent, clear and complete. The specs are often left to the end of the latter half of the construction-documents production. A lot can slip through the cracks.

Design Development is the most important stage of design for the owner to cross-check the coordination and potential for conflicts in the consultants' work.

Independent, knowledgeable and objective review at several intermediate stages of Design Development is the only way an owner can be assured that the drawings are consistent and coordinated. That the specifications have been brought along consistently with the design. And that their evolving costs are still in compliance with the budget. That's what the B/BA is designed to accomplish.

Design Development is where the Design Production Plan required from the architect—with its definitive, production-based payment schedule and its intermediate checkpoints and constructability analyses—will really pay off.

Once the owner accepts the Design Development Documents, the design process enters the Construction Documents phase. This is sometimes called the *Working Drawings* phase, but this is an inadequate term for producing these final documents. It is misleading because it gives short shrift to the need for the specifications, which should be advanced concurrently with the drawings through all phases of design.

Once the owner authorizes the start of Construction Documents, the owner's control over the adequacy and accuracy of the documents and their compliance with the Program will have passed downstream. The design team is now rushing to put the details to the design decisions already made. Changes now will be disruptive and probably costly, will likely delay the project, and will disturb any reasonable confidence in the accuracy of the documents.

Construction Documents

If the architect has been required to produce a Design Production Plan and its commitments have been met through the Design Development phase, the production of the Construction Documents (CDs) should be straightforward. The most important consideration for the CDs is to ensure that the details and specifications are consistent, that adequate cross-sections have been cut where they are required, and that realistic cost estimates and contract duration have been established.

The Easy Ones

It's typical of incompetent design work that the architect and other members of the design team will cut sections through the building where interpretation is straightforward, but will avoid cutting sections and showing details where it is difficult to conceptualize how various surfaces meet.

When the contractor discovers that the surfaces don't meet, or that the dimensions don't work out, the cost of the discrepancies will find their way back to the owner (as in the City Hall example on page 56).

The architect and his design team should, by this time, be believers that the owner is determined to hold them accountable for the documents they will deliver. They should know that their performances will be held to the same warranty standard those documents will deliver to the contractors.

By the start of CDs, the owner should have a well-detailed pre-construction schedule and cost estimate in hand, as the products of the independent Buildability/Budget Analyses he has undertaken. The AIA A201 or its equivalent will have gone up the chimney with the ashes of the AIA B141 (the standard architect's agreement).

By the completion of CDs, the owner should be satisfied that the project has been "built on paper" to the maximum extent it is possible to second-guess the construction trades.

Bidding

Some public agencies believe that all they can measure is the bonding capacity of the bidders. More progressive agencies have discovered the benefits of establishing more meaningful standards for evaluating bids.

9

It's impossible to give this subject its full due right here because of the many different formats of construction contracts. It is important to say, though, that there are ways to ensure responsive bids on every construction contract, even in the public sectors of building construction. An afternoon with the past couple years' issues of the *ENR (Engineering News Record)*, or review of some old specifications and bid documents (available from most contractors, if you only ask) will suggest a considerable number of them.

Construction

Among the owner's concerns is how he can obtain objective, independent overview of performances—both the architect's and the contractors'—during construction.

Put the requirement for controls into the contract. Leave their interpretation and acceptance to the incompetent or indifferent architect or construction manager and the owner has just handed the contractor weapons with which he is much more competent than most architects or construction managers.

Many projects have run into trouble because the owner was led to believe that just putting tough requirements into the construction specifications was enough to ensure responsive performances during construction.

It often goes just the other way around if the owner's representatives are not competent. Putting tough requirements on the contractor, in the hopes that the architect, or even the "construction manager," will know how to implement them, has doomed more projects to disputes than it has helped.

Every requirement—whether for scheduling, payments, changes, time extensions, etc.—is a broadaxe with two edges.

By putting time and cost controls into a contract and then leaving their implementation to the incompetent or indifferent, the owner hands the contractor weapons with which he has much more facility than most architects or "construction managers."

A contractor who, for example, is competent with detailed CPM scheduling can build a case for contract extensions into the original schedule. Acceptance by an inept contract administrator will hand the contractor a license to steal.

The contractor is fully capable, because of his direct control over the performance of all the parties on the jobsite, of thoroughly confusing the project documentation if the owner does not have an independent record and someone available to analyze it objectively.

In Parts 8 and 9, we'll start to put together the procedures it takes for an owner to make effective use of these tools to measure,

124

document and effectively motivate project performances. One of the tools we will recommend for the owner to establish from the start of design, and to continue through construction, is a scheduling system through which the owner can independently, and in parallel with the consultants and contractors, process schedule information, so he can check their reports and their progress.

> In 1969, the author assisted a federal agency to develop procedures to process, in the same standardized systems all contractors were required to use, the reports, schedules, payments, logs and change orders on all of its projects nationwide.
>
> When not administered by idiots—which they sometimes have been—these procedures have contributed significantly to the reasonable negotiation of construction disputes.

Payments to the Design Team During Design

We have commented, several times, that "75% Design Development drawings" is meaningless. One of the important, but difficult, matters to resolve, as a component of the Design Production Plan and the architect's agreement, is how production will be measured and the architect will be paid.

An owner should not compromise these considerations, once defined in his Plan. If an owner finds himself already working with the architect before resolving a definitive payment schedule, he will find himself making continuing concessions to the architect's assertions of progress. Design production and payment are matters that need to be resolved *before* the initial interview—when the RFQ is being prepared.

> ### *"This Guy Scares Me"*
>
> In a talk at the Engineers' Club one night, I was making many of these same points about measuring the production of construction documents and documenting the performance of the design consultants during construction, when an architect in the audience turned to the person sitting next to him—not knowing he was one of our consultants—and said: *"This guy scares me."* The architect was on the design team of the project in the story on page 126.

9

What the owner needs to do is to project the number of checkpoints he expects to require in each phase (this number can be adjusted as the Program and Schematics later dictate). He can then incorporate into the architect's agreement a basis for payment correlated to these checkpoints.

The architect should be required to prepare a schedule to meet each checkpoint, and define the labor-hours he estimates for himself and the design team to deliver these documents. The architect should then be required to prepare a payment schedule based upon the labor and costs associated with each checkpoint.

There's nothing mysterious here, except possibly that most architects have never taken the trouble to understand definitive, resource-loaded scheduling. It's exactly what contractors are required to do on every CPM project.

However the checkpoints are adjusted later, the amount negotiated as the architect's fee will establish the limits for the costs allocated.

This exercise will likely prove an educational process for the owner as well as for the design consultants, since the scheduling of design effort is something the architect asserts cannot be done.

"We Run a Project with Style"

The new high-rise nursing tower for the public hospital facility was to cost $55 million. For a year during planning and design, I had been part of the management team. With limited opportunity to monitor design production because the contract with the architect had been previously executed, I continued to urge, unsuccessfully, that [1] constructability reviews (what we're calling **B/BAs** in this book) be undertaken progressively; and that [2] project documentation requirements be incorporated into the construction documents that would track the performances of both the competitive-bid contractor and the design consultants.

The project architect continued to insist that these measures weren't necessary and would only interfere with the relationships during construction because *"We don't need them… we run a project with style."* As the project went to public bid, I left the management team.

The project ultimately cost $85 million.

126

"Design is a creative process—it can't be scheduled."

But it can be, and the owner should insist upon it.

Retention is a time-honored way for the owner to hold back monies from a contractor against potential problems. The retention of design fees (cries of outrage!) can serve a similar function in relation to the quality, timeliness and production of the architect's products.

What and How Should You Pay for Design Services?

There is a general perception that design fees are usually based upon a percentage of the construction cost. It only takes about three seconds to realize that this is a counterproductive basis for establishing the architect's fee. Knowledgeable business owners have long recognized this and typically negotiate a fee based upon a defined scope of services.

Many public agencies, however, continue to use a sliding percentage scale for evaluating what the architect should be paid. Since in the public sector fees typically slide toward the established maximum, this has invited heavy lobbying by the AIA (it is particularly notable in departments of education).

If you think about construction for a moment, however, you will realize that it should be as straightforward—more so, in fact—to establish the architect's fee on defined services and scopes of work. The architect will have substantially fewer subconsultants than the contractor will subcontractors. The owner is required to produce a preliminary program that the architect will work with him to refine. At that point, the architect should have at least as definitive a basis to commit to a scope of work as the contractor will have with his documents.

Before the start of Schematic Design, then, it should be possible for the architect to develop—and the owner committed to the success of his project should insist upon—a design-production schedule that defines the roles of all the major subconsultants, that commits to a firm delivery schedule for intermediate production of documents (for the progressive B/BAs), plus labor hours and a rate-based budget for each of those delivery points (milestones).

9

If the total of this proposed budget exceeds the 6%, 8%, 12% or other parameter that the owner considers an appropriate maximum for the architect's services, there will be, at least, the basis for evaluating where, how and if the fee should be agreed to, or negotiated further.

It's not incidental to consider: If the architect can't do this, how well will he or she represent the owner if problems become the order of the day during construction?

Why Does Logic-Based, Detailed Scheduling Apply to Contractors— But Not to Design Consultants?

You've heard *the architect's* answer:

"Creativity can't be scheduled."

So the owner asks:

But once I authorize you to start Schematics, don't you define the assignments of the other design consultants?

To which the architect replies:

"But it's still a creative effort that requires coordinating consultants from different disciplines, all of whom have different perspectives and workloads. We can't schedule them."

So the owner sets up the rules:

That's exactly why you should schedule them. Isn't it true that if just one of them slips, it could delay the construction documents? And couldn't it also lead to failures in the drawing coordination?

And the architect's, final, frustrated response is:

"But I'll have to charge more if you want me to develop a detailed schedule and keep it up."

To which the owner *should* answer:

Not if you want this contract. But if you don't have the systems or the disciplines to prepare a schedule, we (the owner and guide) will work with you to help with the scheduling technique, and we'll even provide the system to process information. What you will have to do is provide a definitive, labor-hour-loaded schedule that explains

128

[1] that the fee we've negotiated will produce what you've promised you'll produce, on schedule, and [2] assurances that your other consultants have agreed to it.

You can expect to incorporate into your schedule periodic meetings with me (us) and all of your major consultants to discuss the status of the documents, and the results of our periodic Buildability/Budget Analyses.

The owner may have to hold the architect's hand (with the guide at his side) through the development of a logic-based schedule that assures all the bases are being covered (even though that will remain the architect's duty), but it's not something from which the owner should back down. The implications for costs and time are too great.

It usually requires only one or two successive Buildability/Budget Analyses to demonstrate why requiring the architect to work within a firm timeframe can be so important.

New School-Old Story

The new grammar school in the Southwest was to be built on a virgin site. The community and the board had laid out a well-defined program with separate classroom structures, and a combination gymnasium, arts-theater hall and auditorium that would serve the community.

The community was pleased with the scheme that had been presented by an architectural firm whose offices were 450 miles away. With the contract authorized, the district had only periodic meetings with the architect. The board accepted the architect's assurance that the project could be constructed for its budget of $4.3 million.

A reputable, local contractor provided a low bid of $6.1 million. Asked to assist, the contractor was able only to reduce the cost by $200,000 on the basis of the current design. All bids were rejected, and the architect assured the district that it would find ways to reduce the cost.

It took the architect eight months to provide revised plans for which the original low bidder, as a courtesy to the district, estimated a new cost of $5.7 million (still $1.4 million over budget).

The district retained an independent, construction-knowledgeable consultant to work the design over with the architect. The consultant found that:

1. The exterior siding of all buildings required the interfacing of four separate trades, increasing the costs substantially—and increasing the potential for water penetration and maintenance.

2. The structural framing, the walkways connecting the buildings and roof systems were all similarly complex, with exotic connections. All could be replicated with simpler systems without negatively affecting either the aesthetics or functions of the facility. They would, in fact, improve its operating and maintenance costs.

The project was redesigned and was finally re-bid and constructed for $4.6 million. Had all of these changes been incorporated into the original design, two years of cost escalation of 5.5% per year could have been avoided, suggesting that the original project, effectively designed, could likely have been built for about $4.2 million.

5

THE LAW

By definition a Contract is intended to contribute benefits to the parties that sign it.

Since it also invites the intrusion of *the law,* it's essential for the parties to a Contract to be aware of how the law and those who practice the law can entirely frustrate the purposes all parties to the Contract had in mind when they came together.

INVITATIONS YOU DIDN'T KNOW YOU ISSUED

It's a good bet that your construction documents will not be en-graved. But they will issue an R.S.V.P. to every contractor that will work with them.

Unless you've ensured that they're clear, consistent, coordinated, unambiguous and buildable within identifiable cost limits, the R.S.V.P. responses you get can destroy both your equanimity and your budget.

The responses, of course, can come in the form of change orders or delays. In their worst form they can be contained in a com-plaint that signals the birth of a construction claim.

As you might expect, the birth of a claim can also spawn some unsatisfactory responses from those you retained to help you avoid exactly these circumstances. Following are some of the diffident, postpartum responses that have been heard from own-ers' representatives after the birth of a construction claim:

Codes? The project architect:

> "We rely on our subconsultants to check the codes. Anyway, it's the contractor's responsibility to see that he complies with all the codes. We can't be responsible for what the fire mar-shal does after he comes on the job."

Costs? The architect's consultants:

> "Well, we can't be sure what the contractors are going to bid. We rely on our outside cost estimators to tell us where we might have to make some adjustments. Yes, they only do component estimates, but they're usually pretty good."

Schedules? The architect's spec writer:

> "We usually insert the same spec the government uses. They have a critical-path specification that seems to cover the kitchen sink. No, our office doesn't use the CPM method, but we ex-

133

10

pect the contractor should know how to use it. Well, no, we wouldn't rely on it for measuring delays. Contractors don't seem to keep up their schedules anyway. We pretty much respond to problems as they come up."

Reports? The construction manager:

"We rely on the contractor to keep the meeting reports. We can always correct them if we find errors. Well, yes, sometimes we set up an agenda, but usually we leave that to the contractor. No, we don't check manpower. We have to trust the contractor to do that since he's on the job every day. Besides, we ask him for certified payrolls at the end of the project in case there's an issue of overcharging. But we produce plenty of paper anyway—that's why the owner wants us here."

Logs? All the folks above:

No, we don't keep logs of RFIs. We could if you wanted us to—of course, it would take some man-hours, so we'd have to adjust our fees.

Design Errors? The Senior Architect-Partner:

Well, if the contractor raises an issue of design, of course we try first to resolve it with him at no extra cost. But if he insists, then we'll have to prepare a change order for your signature. And of course, we might have to notify our E&O carrier if it amounts to very much. We can't be expected to produce perfect drawings. And the contractor's required to find errors in the drawings, so he bears most of the responsibility for bringing them to our attention.

Disputes? The owner's attorney:

We usually modify the disputes clauses to keep up with what's going on. Right now mediation's in—and it's supposed to be more effective in the long run than arbitration—but then, you never know. Well, we don't have much to do with the project documentation until a claim is filed, but I guess it would be a good thing for the owner to make sure that he gets good records. That's pretty much for the CM to recommend.

And All the Other Processes:

You can pretty much expect to get a lot of self-serving answers—
and some pretty vacant looks—in response to other challenging
questions about:

- RFIs (Requests for Information and related logs)
- CPFF (C+FF, Cost-plus-fixed-fee contracts)
- Force-Account
- Design-Build
- "75% design development"
- CPM (the critical path method for scheduling)
- "Periodic Observation" (*who decides what the period is?*)
- Redundancies in the scopes of design, construction and CM agreements
- Excusable versus non-excusable delays
- Compensable versus noncompensable delay impacts
- What *are* impacts?
- The relative merits of mediation, arbitration, trial or negotiation
- The *differences* between delay and disruption damages
- Construction estimates versus budget estimates
- Distinctions between cost reports and auditing of estimates and construction budgets
- Segregated-Prime Contracting—pros and cons
- Partnering
- Documents essential to prove, or respond, to a construction claim
- What form of ADR is best for *this* project?

Throughout the following chapters we'll be discussing how these
phrases, acronyms, terms and concepts relate to, or interfere with
an owner's ability to deliver a successful project.

By the time we've visited these ideas in several different contexts, you just might be dangerous enough to ask your own challenging questions of anyone who offers you laconic answers and
empty promises about their contributions to your project.

Every promise and easy answer you nail down will eliminate
another unwelcome stranger from your guest list.

RESPONDING TO YOUR INVITATION—
A war story without an ending—for now

While preparing shop drawings for the steel frame, the fabricator found dimensional errors. He advised the structural engineer of them with his submittal.

The structural engineer returned the submittals, incorporating changes in the location of two structural-steel columns. The architect then issued an ASI (Architect's Supplemental Instruction) to the general contractor advising in a memo (with no details) that the columns should be relocated, but that neither additional compensation nor time would be provided.

The general contractor immediately advised the architect that relocating the columns would require additional changes in [1] the steel roof framing, (already in fabrication), [2] the lighting and ductwork suspension from the roof structure, [3] layouts for electrical fixtures and transitions in the ductwork, [4] the layout of roofing panels, and [5] other equipment suspended from the roof framing. The contractor advised that there would be additional costs and at least 5 weeks' delay for engineering, submittals and fabrication modifications.

The contractor also noted that, because the building was part of a school project, both the contract and state regulations required state approval of structural and other life-safety changes and that formal change orders were required for all changes in the work.

Both the owner and architect responded that, since the project was already over budget, no change orders would be allowed. The architect directed the contractor to perform the work under penalty of being terminated.

The contractor—taking a considerable risk—proceeded with the work under protest, in order to keep the project going. Concurrently, he presented a change proposal, which the architect rejected—as he did all subsequent change proposals by the contractor.

The project was the proverbial iceberg. Design errors continued to surface until the contractors had accumulated over $400,000 in proposed, unrecognized change orders. Under the terms of the contract, the general contractor brought a breach of warranty claim on the basis of the owner's failure to pay for work performed. The owner's response was to terminate the contractor.

With the cat out of the bag, the festivities had only just begun.

11
THE LAW AND LAWYERS

Introduction

Slabs of black diorite, buried for almost 4,000 years, were unearthed in Mesopotamia (today's Iraq) in 1901. These tablets are engraved with elements of the *Code Hammurabi*.

Hammurabi is credited with being the first ruler to apply rules of order to the chaos generated by humanity's proclivity to extract personal vengeance for perceived wrongs. He largely succeeded in his time. He also succeeded in creating a new occupation: *the law.*

The practitioners of that occupation—the legislators, the courts, the lawyers of our Western civilization—have had 4,000 years to work on Hammurabi's premise that justice would arrive with order.

Those practitioners, and the lobbyists who influence them, have had their own success in turning the law into a self-perpetuating industry. In the process, they have repositioned the law from serving order, reality and justice to serving, in large measure, themselves.

They have done this by inventing a private society with its own language and almost-secret ceremonies. Legalese, conferences between opposing attorneys (without clients present), conferences in chambers, the imbalances that accrue in different legal venues (more about this in chapter 13), secret rules and rubrics, all militate in the direction of excluding participation by anyone not part of that closed society. Rules often prevail at a cost to justice. Those who speak the language of law use measured obfuscation to perpetuate the dependence of clients.

In this, the practitioners and influencers of the law are abetted by all of us. We skirt it. We try to avoid it. We work it to our own interests. In the process, we make it virtually impossible to clean up the messes we've made.

Today, in that part of the law that was once directed to honoring men's promises to one another—Contract Law—Hammurabi's original purpose of applying order and justice to human relationships is served almost incidentally, if at all, and often only accidentally.

11

But The Law Is Not the Monolith
You May Think It Is

It is useful, though, to know that Contract Law is less a monolithic structure than we laymen have been led to believe.

Until the 20th century—in fact, from the time of the American Revolution—business continued to have great influence over the direction of contract laws. Even in our colonial period, it was clear that private and commercial interests on this continent would not tolerate the intricate mechanics that had evolved in English Law. Attempts to introduce convoluted pleadings into American Law were dumped into the harbor along with English tea. American Contract Law evolved along more pragmatic lines to meet the needs of American business.

Judges were often laymen, chosen for their common sense and ability to see through legal rules to the heart of the matters behind contracts.

By the early 20th century, however, practitioners of the law had closed ranks. Its practice required admission to "the profession". In 1908 (only seven years after Hammurabi's tablets were rediscovered), the American Bar Association adopted a professional code of ethics. Lay judges became a historical footnote.

By the early '20s, Harvard Law School's Samuel Williston, following the legal philosophy of the *conceptualists* (absolute rules could be set down for every tenet of the law), developed (in multiple volumes) the Holy Grail of Contract Law. Anyone who attended law school through the '60s was required to apply Williston meticulously to examinations.

But others in the profession, following the counterpoint of *realism* argued that Contract Law should be motivated not by fixed rules, but by the realities to be found in particular situations and environments.

The debates that ensued, along with many others that pit "liberal" versus "conservative," "individual" versus "societal," ad nauseam, continue today within the profession. Each state has enacted civil regulations unique to its commerce and varying dispositions. The application of the rules of Contract Law continue to be constantly modified, influenced by the clubbishness of the profession and by the dispositions peculiar to local counties, local courts and constantly changing political influences. The law has produced what comes down to *"the local rules"* (each court has its own).

How Does the Territory of the Law Relate to the Owner's Project?

This is one of the earliest questions the owner has to consider as he sets up his Project Plan.

Of all the territories from which the owner must select the "essential strangers" to assist with his or her project, the law is the strangest.

Of all the unfamiliar languages spoken in those territories, the law is the most foreign. Of all the rules that will apply to the interrelationships of the participants, the rules of law are the most arcane.

Since involvement or entanglement with the law—in the form of contracts, codes, regulations, standards and precedents—is inescapable, however, the owner has to choose from among all the "strangers" and "other strangers" out in that territory.

The owner will need the assistance of a *fiduciary[1]* attorney to interpret how the law and the local courts are likely to view the requirements in his contracts.

If that were all an owner had to consider, it might be enough to find a construction-knowledgeable attorney (oxymoron?) familiar with the local rules. But few attorneys are familiar with what happens in design offices, contractors' offices or jobshacks. Few really know what the strangers from those other territories need to do behind those doors to make the owner's project successful. Those are the performances that need to be measured and comprehensively documented.

The requirements, measurements and documentation of those performances are defined in the scopes of work and in the specifications that are parts of those agreements. Development of those scopes and specifications requires the perspectives of someone who has experienced what happens behind those closed doors.

> *It's one thing for an attorney to have seen the aftermath—the results of failures to perform adequately behind those closed doors. It's entirely another to have been behind those doors long enough to know what happens there.*

---

[1] *Your attorney may tell you that fiduciary in relation to your relationship with your attorney is a redundancy. You might want to discuss with him or her the Onsite/Insight newsletter included in this chapter.*

If you give serious consideration to the recommendations in this book you will need to bring together at about the same time, both the attorney you will rely upon for the duration of your project and your guide.

THE LAW OF CONTRACTS

Contracts and Consideration

The glue that holds a contract together is known in the law as *consideration*. *Black's Law Dictionary* defines consideration as:

> *The cause, motive, price or impelling influence which induces a contracting party to enter into a contract. The reason or material cause of a contract.*

As you consider the concepts, the ideas, the requirements and the measuring tools that will go into planning your project, you probably won't be thinking much about the skeins of rules, regulations, codes, standards of performance and legal precedents that will later interpret the applicability of those concepts.

It's similar to how you might approach creating a landscape painting with your project as the centerpiece. You assemble your brushes, your paints, your imagination, your images of your final project. Although it's right there in front of you, you don't think much about the canvas—the fabric that will hold all your images together. That's *consideration*.

That fabric, for your project, consists of the promises you will make, and expect others to make, to maintain order in the development of your project. *Consideration* is the exchange of those promises.

The premise is that the parties to the contract willingly agree to trade one valuable thing for another, with each party believing they'll receive what they've been promised. Since, in construction, those promises can become pretty complex and extensive, the words used to define them need to be carefully crafted.

Crafting and interpreting complex promises and counter-promises is what attorneys study in Contract Law. Contract Law, however, does not cover how to measure and document the fulfillment of those promises[2].

---

[2] *Some law schools, however, compensate for this shortcoming by requiring their graduates to undergo a lobotomy-ego-adjustment to compensate for any sense of their own limitations. That's why many of us have been led to believe this shortcoming doesn't exist.*

To help you define the means—the tools—to measure and document those performances, you will need the combined talents of your guide and an attorney committed to the ideas you decided to select from the recommendations in this book.

You will need your guide to describe and define the conditions in which those tools will be implemented. You will need your attorney to make the requirements to use those tools enforceable—the tools that are designed to protect your project from the Aftermath.

The Aftermath

The Aftermath (not a magic buzzword, just another shorthand) is what happens after the math doesn't add up. When somebody asserts that somebody else didn't keep his or her promises. When the performance doesn't meet the applicable standards. When the performance delivered doesn't match the money paid.

That's when you'll need the documentation.

If the contracts you authorize don't contain provisions for delivery to you of contemporaneous measurements of progress and performance, your chance of obtaining these records later are between slim and none.

Once things hit the fan and you try to obtain these records, you won't believe the legal difficulties and the costs of trying to get them. If you do get them, you probably won't recognize the project as the same one that was reported to you each month.

The Aftermath may or may not resemble Reality. Assertions aren't the same as proof—sometimes they're just balloons floated to test the atmosphere. But if you can't deflate those balloons right away with facts, they can stay afloat a very long time. And then they can be very costly to shoot down.

That's why the Plan you develop needs to emphasize concurrent, continuing and comprehensive measurement of the performances of all the firms you retain to help develop your project. That includes your "construction manager[3]."

---

[3] You can decide after chapter 16 whether your own "construction manager" is less a stranger than other strangers.

11

Our purpose in this chapter and in chapter 13 is to help you determine the level of support you will require from your lawyer. In order to do that, we will visit in chapter 13 some of the legal arenas in which your own performance and those of the parties you hire might be tested.

The discussions that follow are just suggestions of the kinds of discussions you might want to consider having with your guide, and with one or more attorneys you trust. Your disposition in this regard might be very much like the one we suggested in regard to architects: check with others whom your prospective attorney has served—or screwed[4].

The Attorney You Need

Before you discussed your project with anyone, you had a head full of images and ideas. Ideas stimulated by personal dreams, by school curricula, patient loads, marketing plans, income proformas, library populations, or production rates.

Even before those ideas begin to take form in your Preliminary Program, they have been surrounded. Surrounded by skeins of state laws, building codes, local requirements, professional standards, warranties, legal precedents, and federal regulations. You haven't seen them yet, but they're out there waiting to snag your project.

Before your project is complete, your original ideas will become entangled with more contracts, funding agreements, specifications, bid documents, contract modifications, agreements—and disagreements. The implications of the laws that govern these transactions are invisible to you—even if some of the documents that contain them are right in front of you.

In the lawyer-laden culture of the U.S., Hammurabi's original contributions to justice and order are so overlaid and entangled they contribute as much to abuses of order and justice as to their realization[5].

> You ask your attorney to review the Agreement you've been offered by the architect. Whether or not you also provide the essential companions to that Agreement—the general and

---

[4] *An interesting book on this subject is Donald E. deKieffer's book,* How Lawyers Screw Their Clients.
[5] *To wit: Philip Howard's,* The Death of Common Sense.

special conditions governing the architect's performance—it's a fair presumption that your attorney has only limited experience with what the promises in the Agreement portend.

In fact—whether or not your attorney even asks for those general and special requirements and scope of work is a fair test of whether he has the perceptions and skill you will need to have available.

Unless your attorney is uniquely knowledgeable in construction litigation, he will review the disputes clauses, may look over the insurance clauses, is likely to discuss your general view of the fees you've agreed to pay, and bless the Agreement for your signature.

Even if he is a "knowledgeable construction attorney" (Sam Johnson might suggest you look for evidence of the lobotomy here), he will probably have little to contribute in terms of the means and methods you will need in order to measure how well the architect or the contractors keep their promises.

On Your Last Project, Did Your Attorney...?

Assume for the moment that you've been involved in the development of a building project before (many of you probably have). Before you signed the first contract on your project, did your personal, corporate, city, country or agency attorney:

- Review with you how the various relationships that would develop on your projects would be measured, controlled and documented under the contracts you were about to sign?

- Review the coverage you were getting with the architect's E&O policies, or the bonds you were paying for? Or the conditions under which you might have to call upon them? Or the prospect that some of the coverages you were buying might be used up by the insurer in the defense of the insurer's principal ("wasting policies")?

- Describe to you the warranties you would be making to contractors as you produced your documents for negotiation or bidding?

- Discuss with you what is involved in asserting and proving—or disproving—the merits of a contract dispute?

- Discuss the documents you will need in your files in order to negotiate effectively or to defend an assertion of contract damages?

- Discuss the distinctions between excusable, non-excusable, compensable and non-compensable delays during construction?

- Discuss whether those agreements contained adequate protection from the liabilities of deficient design documents, or from deficient design administration during construction?

- Review with you the options you might have for resolution of disputes (trial, arbitration, mediation, ADR, discovery, etc.)? Did he explain the documents you will need, and how you will proceed to get them, in the event of a claim?

- Discuss with you the options you might have to obtain objective, conflict-free representation of your concerns for delivery of a successful project?

- Discuss with you the options you might have with regard to various contract formats for relating the performances of the design consultants and contractors during *both* design and construction. For example, C+FF, Design/Build or Segregated Prime? (You'll learn of some of these from your architect, but you should have some independent perspectives beforehand.)

- Discuss your obligations to your design consultants, to the contractors, and to all the other parties to whom you will be making warranties, explicit or implied, with your contracts?

If your attorney didn't initiate these discussions, it's a fair bet it's [1] because he's directed more toward dealing with bodies delivered to the operating table than with preventative medicine, and/or [2] he lacks experience with the means and the techniques necessary to protect your budget, to ensure the buildability of your construction documents, or to acquire and work with the documents you need.

The training your attorney received in Contract Law dealt principally with [1] setting up the promises being exchanged: the tit-for-tat called *consideration,* and with [2] what to do when those promises are broken.

What's missing in his experience is what happens in between—the performances required to fulfill those promises.

Where *Do* Attorneys Fit into the Project Development Process?

Conflict and confrontation have always been a part of construction.

Construction claims arise—torts[6] aside—because one or more parties assert that one or more other parties didn't perform under the terms and conditions of a contract they executed.

You might think it would be in the interest of the parties to a contract to write their agreement in words not only they, but a bright high school student could understand.

The opposing words of two lawyers, however, demonstrate why that is seldom so:

> Thomas More, writing in 1516, said, *"(so) every man may know his duty... the plainest and most obvious sense of the words is that which must be put on them."* (It may be significant that Sir Thomas lost his head because he was so clear about this.)

> But another attorney in 1951 wrote: *"Law... ought to be unintelligible because it ought to be in words... and words are utterly inadequate to deal with the fantastically multiform occasions which come up in human life...."* (As Philip Howard makes clear in his book, this is the prevailing viewpoint.)

So construction is irretrievably tied to the business of the law, and potentially to the practice of obfuscation. Lawyers are disposed to keep it that way. Woe to the owner who would sign a contract without his guide and his construction-knowledgeable-interpreter-attorney at his side.

And even then, what assurances? There never has been a contract perfectly written, or perfectly implemented.

In the last analysis—and the *raison d'etre* of this book—the best protection an owner can have is to understand the requirements going in, and to insist that they be incorporated to the highest degree possible in his construction documents. Then if someone has failed to perform, the owner will at least be equipped with some documentary evidence of responsibility.

---

[6] *A tort is an action brought to remedy an asserted injury other than a breach of contract. Although some of the injuries that can arise during construction meet this test, we're only considering in this book those assertions by one party that another failed to perform under the terms of a contract.*

ONSITE INSIGHT©

CANDID COMMENTARY ON CONSTRUCTION, COSTS AND DISPUTES RESOLUTION

DO YOU LOVE ME OR HATE ME? WHAT SHOULD I DO ABOUT IT?

Some of the country's busiest construction attorneys convened at the 9th Annual CONSTRUCTION SUPERCONFERENCE in San Francisco in November 94.

This year a certain paranoia was evident as result of the jokes and criticism* heaped on the legal profession in recent years. It became the theme of the opening session: "Let's explore the Love-Hate relationship between attorneys and their clients."

A fascinating set of conditions is set up when someone asks, *"Do you love me, or hate me?"* The person asking the question usually has several tiered defenses already in place to protect their ego, or their id.

Paranoia is to be distinguished from true humility in that once the perceived threat is removed, it is not always replaced by a clearer perception of reality. It could be productive to measure the effects of the discussion upon the practices of the legal profession in the next few years.

Each of us has our own opinion regarding how much of the criticism and paranoia is deserved, so review of the issues raised in the opening session can be instructive for those of us who occasionally work with attorneys, or require legal assistance.

THE GENERAL FORUM

The opening forum was moderated by a panel of attorneys, contractors and corporate clients.

It quickly became evident that what was being talked about was not *love*, but money ... two subjects for which the lines of demarcation aren't all that clear anyway.

* Criticism that has been directed at The Law in general in Philip Howard's recent bestseller *The Death of Common Sense.*

Owners and contractors believed they could be better served if their concerns were addressed [1] through more direct contact with their

---

THE HONEST THING...

It is said that Diogenes the Philosopher walked through the city with a lamp, looking for an honest man....

"Larry, how's that construction case you've been trying to mediate the past few months?"

"I've been doing more research on it, John, and I've come to realize the contractor's been right all along. He really was abused by the architect and the owner. He followed the contract. They not only disregarded it, but thumbed their noses at State regulations in the process. Now his own surety's turned on him. But when you take an objective look at the facts, the deep pockets have ganged up to put him out of business rather than do the honest thing."

"Larry, I've been getting calls from the surety and from the architect's E&O carrier. You know how much legal business we've done with them."

"You mean the firm wants it to go the surety's way? You want to see more pressure on the contractor to settle for as little as possible?"

"Larry, you know better than I do how important your 98% settlement record is to you... I'm just telling you there's no sense being odd-man-out, even if you think the contractor's being screwed. Think about it."

So Larry did.

And Diogenes, who had paused for a moment, picked up his lamp and moved on.

THIS NEWSLETTER IS PRODUCED IN THE NAPA VALLEY BY **BOYD**

THE LAW AND LAWYERS

principal attorney; [2] with well defined plans, schedules and budgets for legal services; [3] with less dependence upon, and substantially fewer hours of paralegals and junior attorneys, whose perceptions of issues seldom contributed benefits to the client (but provided substantial sources of billings for the attorney); and [4] through more effective counseling *in advance* of problems, to reduce dependence upon the legal process altogether.

As the discussion evolved, the following issues got some shrift:

- The need for attorneys to shift the emphasis of their services from litigation to counseling.

- The need for attorneys to participate in partnering sessions.

- The need for more candor between attorneys and their clients; and with all parties on the project. It was acknowledged that candor can present problems for protecting privileged information, but the owners and contractors on the panel were firmly behind the need to find ways to effect it in any case.

- The distinct differences between public and private projects in relation to contracts, partnering, claims, quality of construction documents, and the openness of all parties to directness and candor.

- The perspective that there are "junk-yard dog" contractors and attorneys; and that they sometimes find it necessary to pair up when the owner demonstrates the same qualities.

- A pervasive, almost haunting hankering for a return to the traditional values of honesty, integrity and candor in legal relationships, and in pursuing settlements short of resorting to litigation.

Some relevant ideas that got no shrift in the discussion, but should probably be explored by the rest of us:

- Whether the ideas of more candor, "partnering" and integrity raise other questions about the fiduciary representation we have expected from the profession; and whether it's the client or the attorney who requires the greater re-education in traditional values.

- The failure to address anywhere in the discussion the fact that the most pervasive sources of construction disputes are poorly defined project planning and programs, ineffective contract administration and just plain bad design documents; and that most attorneys have very limited perspectives about how to improve any of these.

In future O-Is we'll be discussing these ideas in the most direct, candid, no-holds-barred way possible. Let's hear from you.

JOE BOYD, IN THE NAPA VALLEY

Who is O-I

O-I is being published in the prospect of sharing the experiences a handful of us have had in the construction industry over the past 40 years. Those years have seen a steep decline in the professionalism and productivity of management, design and construction.

Over this time the increase in claims, the deteriorating quality of construction documents, the inability to communicate, the increase in professional paranoia, the loss of management expertise, the proliferation of attorneys, have all taken the fun from an industry that once provided great satisfaction.

So we've decided it's time to say-it-like-it-is.

We believe there are solutions – simple ones – solutions that require reassertion of and perseverance with the basic values that made the industry productive and satisfying 30 – 40 years ago.

Through O-I we hope to share our experiences with those who've had the same thoughts; and those who might benefit from them.

SEND US YOUR WAR STORY

Tell us about an experience you've had – good or bad.
E-Mail: bit@splitrockpub.com
FAX: 707-226-1438
Snail Mail: Split Rock Publishing Co.
P.O. Box 5517
Napa, CA 94581

11

These were the precipitant matters of concern to owners and contractors in the opening session of the Superconference in San Francisco in 1994. This is an annual conference at which the principal participants are some of the more eminent construction attorneys in the U.S. Their clients and members of the construction industry are also present. The subject of the opening panel in 1994 was: *"Why do you (owners and contractors) hate us?"*

Through a two-hour, often candid discussion, punctuated by *mea culpas* from attorneys whose obvious and absolute sincerity has served them well, the panel unanimously agreed with the following conclusions:

- There is a need for construction attorneys to shift the emphasis of their services from litigation to counseling their clients on how best to avoid disputes.

- Attorneys should be actively involved in (but not lead) partnering sessions.

- More candor is called for between attorneys and their clients despite some concern for confidentiality.

- Clients expect from their attorneys more conscientious budgeting of time and firm projections of costs, instead of the open-ended billings they had been receiving for years.

- Owners and contractors both expect more direct input from the principal attorney they believe is handling their case, and less billed time of junior associates and paralegals, inexperienced in the complexities of their case, but nonetheless handling important depositions and document reviews.

- The legal profession should make a concerted effort to reconstitute honesty, integrity and candor in all its relationships with the industry and should emphasize honest and objective settlement of disputes over litigation.

Several panelists and other attendees also made the point that even with these transformations in their relationships with their clients, an effective attorney would still have to match wits with the "junkyard dog" attorneys, who are the bane of the construction industry. It was uniformly agreed that this particular breed is not likely to be eradicated in the foreseeable future.

Competent and conscientious attorneys, before and since this conference, have often commented that the construction industry, in particular, contributes to the livelihood of some of the worst of the legal profession.

That should at least get the attention of an owner who has initiated a construction project believing that truth, justice and fairness will prevail in the event of a dispute.

That Superconference took place at the end of 1994. It would be interesting to know how many of you have witnessed spontaneous conversions within the legal profession.

It would also be interesting to know whether introduction to formal courses in *ethics* in graduate school (at the ripe age of 23-25) has produced any noticeable benefits for either the attorneys or their clients.

The challenge for you as an owner, as with the selection of an architect or a contractor, is to shuck the intimidation the legal profession exudes, and to employ the insights and leverage acquired from your own experiences to identify the qualities you will need in your attorney when those "other strangers" begin to appear.

One effective way to do this is to expand on the list under, "Did Your Attorney..." a couple pages back and insist that the person you're considering explain these and other essential realities to you in a vernacular with which both of you are comfortable. This was one of the other issues raised in that 1994 conference.

Unlike astrophysicists and thermodynamics engineers, who often only have one language to give (Stephen Hawking is an exceptional exception), it's fair to believe that most attorneys can speak Everyman's English. If they couldn't, we wouldn't see so many of them on talk shows. It's not an outsized demand that they explain the ramifications of the contracts you're about to sign in explicit, simple terms.

It's important for your attorney to explain what the contracts he blesses will do, and what they won't do, to protect your budget under the worst of circumstances. You should schedule some extended conversations that include your guide as well as your attorney.

How Common Is It?

Chapters 11, 12, 13 and the stories throughout *BIT* demonstrate a pervasive absurdity about how the law is practiced in relation to construction. Understanding this absurdity will go a long way toward reducing the intimidation we're all supposed to experience when dealing with the law and lawyers.

In *Black's Law Dictionary,* Common Law is described as the law that comprises the body of principles and rules of action... which derive their authority from ancient usages and customs... and do not rely for their authority on the codes and laws legislated by government.

Many states have adopted in their civil codes a statement that says, in effect, that to the extent these customs are not contrary to the laws of either the federal or state government, these accepted rules will have weight in decisions of the courts. This means that the parts of the law which are of general or universal application within the state will apply as they relate to security of persons and property.

When this concept is applied to what we're required to do in the administration of design and construction, you begin to realize that we're dealing with legal interpretations every day when we use the terms and conditions of the contracts we work under. When we apply the specifications and the general and special conditions that are part of those contracts, we're not only working with, but in a sense interpreting the law almost every day on a construction project.

In business law, this puts construction in a unique category. There are few other businesses that invoke issues of *what the contract says* on almost a daily basis.

This makes the assertion that witnesses in a construction matter can't comment about the law in their testimony just one more absurdity that attorneys invoke. We have to deal with it, work with it daily, be controlled by it, *but our opinions can't be considered when push comes to shove.*

The fact of this absurdity, however, is with us. Try to comment in a construction dispute about what you believe the documents say, and you're likely to be shut down by both the opposing attorney and by the court. It's part of the rules of the game attorneys have set up to keep their territory to themselves. They can hardly claim that it's to *protect the integrity of the law.*

Faced with this, it is all the more important for you to find that attorney with whom you can get down-and-dirty to ensure that you understand your legal position, and so that you can get out on the table your judgments about how the law may have been abused in your case.

12

THE GRAY EMINENCES

They're powerful, paranoiac and essentially invisible.

They will have great influence over the life of your project, from the time you execute your first contract through, and even beyond, its completion.

The instant you execute your design contract they're exposed to substantial liabilities, so they will be the Engines of Denial if your project encounters construction disputes.

They represent possibly the highest barrier the owner needs to navigate in negotiating an effective design agreement.

They are the shadowy men and women *Black's Law Dictionary* defines as:

> **Insurer/Surety:** One who undertakes to pay money or to do any other act in event that his principal fails therein.

Their paranoia is natural enough.

Once one participant on a construction project fires a shot, other gun ports open, with multiple parties hoping to be compensated from multiple warchests.

The potential risk the insurer/surety assumed on behalf of its client can quickly become actual risk.

The terms of most contracts in the construction industry require the parties hired by the owner to provide the owner with protection, in the form of insurance or bonds, against the eventuality that the party providing the policy does not perform within the terms of its contract with the owner.

The protection the owner requires may be to protect against loss if the party withdraws its proposal after it has been accepted. This is most commonly known as a Bid Bond. Its protection is generally limited to the amount between the bidder's quotation and the actual cost at which the owner is able to resolve another agreement for the work.

12

The owner may require a performance bond, which provides that if the performing party does not complete the work under its contract for any reason for which the owner is not responsible, the surety will provide the funds, and/or another party to complete the work. It is not uncommon that a surety may challenge whether the reason its principal did not complete the work was due to conditions under the control of the owner—as for instance, incompetent construction documents. Disputes on this kind of issue can leave a project in limbo for months.

In the area of design services, the coverage the owner buys is to insure against the additional costs that can result from defective design. Since the owner's exposure to cost overruns can exceed the architect's total fees (design-defects claims frequently exceed 10% of the construction), many architects are unable to obtain an E&O policy that would adequately protect the owner. Coverage is usually a matter of negotiation between the owner and the architect.

For instance, the owner may want coverage in the amount of $1 million, on a project for which the projected construction cost is $8.5 million. The owner is betting that the impact costs attributable to the design consultant will not raise the actual cost of construction above $9.5 million[1]. The architect may be able to provide a policy that only protects the owner to a limit of $500,000. The owner will have a decision to make.

There are design professionals who promote the idea that the owner should self-insure. They argue that the owner is in a better position to spread the risks over the life of the project, and that it is the owner who receives the principal long-term benefits from the project. Reduced to the simplest of terms this says: *You should let us off the hook because you can afford (and we can't) to cover the cost of the mistakes that might be in our documents.*

To which an owner's response might be: *How is it you believe that you have no liability if you produce defective documents through which I provide warranties to the contractors?*

However the owner approaches the issue of indemnification against defective documents, it will be the owner who will have to make the initiatives. The standard design contract has yet to

---

[1] *See the glossary and chapter 15, however, on wasting policies.*

surface that will volunteer the measurement and documentation of performance, or protection against liabilities, the owner is required to provide to a contractor.

It is also important for the owner to note his own responsibilities to be clear and comprehensive in the definitions in his Program, because subsequent changes, additions and reversals will subvert any warranties from the architect.

In these and the many other eventualities against which an owner may require various types of protection, the insurer/surety is betting that his principal will perform up to the standard of his or her contract. The E&O insurer is, at the least, betting that his principal will not be found legally responsible for conditions that result in losses to the owner on his project.

There are many other kinds of coverage the owner can buy, such as protection against theft, against discovery of unknown conditions, or acts of nature. But we're not concerned with those directly here, since we're discussing principally the conditions an owner can control through development of an effective Project Plan. When the owner is working on the Plan with his guide and attorney, however, these other areas of liability should be considered, since they must be incorporated into the agreements the owner will authorize.

The policy of insurance or bond is provided to protect the owner. Usually the owner pays for it, directly in the form of premiums, or through the terms of the contract with the party he hires.

The coverages are for the benefit of the owner and his project. It is important for the owner to understand, however, that the concerns of the insurer/surety are, in the order of their importance and their priority to the insurer:

- Protecting the insurer's own economic interests,
- Protecting the interests and livelihood of its principals, and
- Insulating the owner against risk.

When an owner requires others to provide an E&O policy or a performance bond, the owner needs to realize that when the

12

Re-defining the Architect

Since the insurer's primary concern is to protect its own economic interests, you have to wonder why the industry doesn't require their principals to meet higher standards for design-document coordination and in areas of management. Or why the insurer doesn't discourage them from commitments they're unlikely to fulfill.

Design creativity is lodged primarily in the right brain, but effective management, cost estimating, scheduling, measurement and documentation are essentially left-brain activities.

If architects and their insurers would re-define design activities and limitations, it would go a long way toward reducing the adversarial relationships architects continue to assert should not exist in the construction industry.

It's a potentially productive field waiting to be tilled.

---

owner makes a call on the coverage to indemnify himself against any but a relatively small loss, the surety will probably take its lead from the Queen:

"No, no!" said the Queen. "Sentence first—verdict afterwards[2]."

The surety will understandably pursue any action that will demonstrate that its principal did no wrong. The surety will usually assign whatever legal support and consulting expertise it deems appropriate to support this position.

If the surety and its principal are unsuccessful in maintaining this position, the next line of defense will be to spread the responsibility—to encourage joining other parties as co-defendants: the subconsultants, the contractors, suppliers and as many others as possible. That may be good business practice (at least if it's done with some justification and without malice), but it doesn't necessarily obtain timely payment for the owner's damages or loss.

The owner who believes that his architect should be held to the same standards of performance as his contractors—that is, that the contractor can be held responsible for *any* failure of performance—can expect very loud objections from the architect's insurer.

[2] *Lewis Carroll:* Alice's Adventures in Wonderland.

They will echo the response you will hear from architects:

Nobody's perfect. You have to expect some problems. We can't be held responsible for all the problems that come up during construction.

Although this assertion is literally correct, its application has often been expanded to include design errors that have contributed substantially more than 10% in cost overruns on projects.

Fiduciary Architect

In 1997, a court found the architect on a $20 million high-rise project to have failed its *fiduciary* duty because it did not follow through on contractor-submitted shop drawings for windows that were found to be deficient.

Black's Dictionary of the law suggests that this would require the architect to meet a standard of care equal to that of a *trustee*: Someone into whose hand the welfare of the project is entrusted by the owner.

Although it's still in doubt whether this ruling will stand the test of appeals, it does reflect an increase of disillusionment with architects' performances.

Recent legal interpretations of the architect's responsibility to the owner for design errors, and for failures to administrate adequately submittals and changes, have supported holding the architect to progressively higher standards of performance.

The owner can expect that the architect's E&O carrier, and the contractor's surety will have great interest in any modifications the owner makes to either of the design or construction contracts. They will have considerable interest in how the law will interpret the contracts their principals sign.

For that reason, among others, it is crucial that the owner, through his construction-knowledgeable attorney, become conversant with the verdicts the law is likely to deliver in relation to his project, his documents and his administrative program—as well as with the role the insurers and sureties will have in the owner's enforcement of the standards of care to which he intends to hold his consultants and contractors.

13

THE LAW'S ARENAS: OPTIONS, AGONIES AND ABSURDITIES

Adversaries?

Professionals and innocents alike assert that construction need not be an adversarial enterprise.

They're right. Either because they've been competent enough to demonstrate the truth of the premise, or naïve enough to believe that just saying it will make it so.

In either case, however, they're talking about design and construction *before* confluence of those two enterprises results in the vortex of a dispute.

Once a formal claim is filed, *adversarial* is a mild term for what happens next.

Trapped in the Arena

You've seen on TV the dangers to life and limb that can erupt at a soccer game.

Have you ever been caught in the traffic jam trying to get out of the ballpark? Remember the hazards of trying to extricate your car from the parking lot? The frustrations on the exit road? The *I wish we had...* as you thought of ways you could have avoided the crush? The crazy drivers, your own and others' impatience as you crept through areas where you'd rather not have to slow down? The dangers that raised the hair on the back of your neck when you could tell other people were out of control?

Put all these experiences together, add just one junkyard dog, and you begin to have an idea of what it can mean to be caught up in a construction claim.

Despite what you may have heard about certain venues being more benign, less costly, shorter or having more productive, reasonable results, any dispute resolution process that goes beyond direct negotiation between the parties will bring frustrations and levels of dissatisfaction you can only anticipate if you've been in the vortex before.

13

> ## Let's Look Around the Ballpark Before You Leave
>
> Look around at the other folks in the parking lot. How many of them do you know? How many do you plan to see at dinner tonight or lunch tomorrow? How many do you expect to see in your office next week? Which ones will haunt your dreams next Thursday morning?
>
> Well, some of them will walk into your office next week and tell you that they have a problem with something that someone who works for you did that they shouldn't have done. Something you know nothing about.
>
> They will tell you about rules you should have followed. That you're responsible for what those folks who work for you did. Rules you didn't know anything about. Activities you didn't know were going on.
>
> They'll bring more of those strangers from the parking lot with them, some of them speaking in strange tongues.
>
> You will have to hire a translator to interpret both their strange languages and the rules they keep repeating. Your translator will begin to have private conferences with those strangers. In just a couple of weeks you won't know what he or those strangers are talking about.
>
> But you will start receiving bills from the translators you've hired. And from the consultants he's recommended you add to your list of strangers.
>
> The disturbing dreams you had that first Thursday morning are returning several times a week.
>
> And all of them—strangers, translators, dreams and *bills*—will be around *for months*.

This applies whether you go to trial, to arbitration, to mediation, through a mini-trial, or one of the myriad Alternative Disputes Resolution (ADR) forums. Once a dispute to which you're a party is filed in the jurisdiction of a court, you will start to meet more "other strangers" than you can believe.

A dispute that's not resolved directly between the parties will be filed with a court before it goes anywhere else, even if your contract calls for arbitration or mediation. Filing means attorneys. That means unrecoverable costs: the costs of pursuing, or defending, a construction claim are often not recoverable.

We will visit some of these venues in this chapter. You will find some of the "other strangers" here, and others in following chapters. These visits will introduce you to more ironies.

These visits have three main purposes: [1] to describe what getting into, and out of, each claim venue entails, [2] to suggest the resources that a participant needs in order to prevail, and [3] to encourage you, the potential owner, to invest every effort you can muster to stay out of *all* of these arenas.

If It Can't Be Negotiated, Where Do You Go?

The first irony is that while your goal, in the event of a dispute, will be to negotiate your way out of experiencing a construction claim, you must decide even before you hire anybody which mechanism for disputes resolution will be best for your project.

If you don't put the same venue of choice in both your design and your construction contract, the courts will do it for you. It is in the nature of most construction disputes that parties to both of these contracts will be involved in the findings of fact and rulings of law. The court of jurisdiction will join all parties involved in a dispute into a single venue so all the issues raised can be tried efficiently. You can lose your venue of choice if they're not the same in all your service contracts.

Jurisdictions

The general rule is that construction contract disputes in which neither party is a government agency will be controlled by the rules of the state superior court in the district or county in which the dispute arises. This is because there is in place a body of state laws, regulations and business codes which apply to your contracts even as you sign them.

If your contracts do not contain a requirement that claims will be tried in some other venue, the dispute will be heard before the appropriate superior court. The rules of that court, operating under state regulations, will apply. And the irony is that the court will order a judicial arbitration ("jud-arb") anyway, or mediation before a court-appointed special master, or some other format it has invented to unload its court calendar.

The courts are so loaded with criminal actions—the same courts hear crimes, contract disputes, torts and white-collar crime—that

13

they will do anything they can to put the less sanguinary affairs into the hands of surrogates.

There is an increasing dissatisfaction, however, among professionals and disputants alike with court-appointed surrogates, because of a developing history of idiocy and self-serving responses to construction disputes.

Our first visit will be to the court trial. Then we'll go on to government hearing boards, and finally to some of the more prominent ADR options.

The Claim Is Filed

Once a claim is filed you can expect that no fewer than four attorneys will become involved: yours, theirs and two insurance attorneys (probably several more before the end of the week). You won't know they're there until you see their names among the addressees on various motions before the court.

Since these attorneys will all be collecting fees for services, at least some will have no practical motivation to cut off the flow of those fees prematurely. This is facilitated by the rules of the game: once represented by an attorney, none of the parties can talk directly: *"You might say something that would negatively affect the outcome of the case."*

Although this prohibition has the benefit of protecting certain interests (including the attorneys'), it also commits all parties to those strange voices and to protracted costs.

It eliminates any potential of negotiating the merits of a dispute.

Trial at Court

One or more parties may insist on a jury hearing and finding. A jury trial may be waived if both parties agree, in which case the matter will be heard by the court sitting alone (i.e., the judge). The judge is usually a pick of the draw, selected by the trial-setting judge of that district, or selected by the disputants from three he makes available.

Regardless of what the attorneys do, and despite all the governing rules and regulations, much will come down to the judge's judgment, disposition and *rulings on the applicable law and motions.*

Opposing parties will bring motions (endless motions) to insist upon document production, to resist document production, to compel testimony, to exclude testimony, to admit evidence or to exclude it. These motions will be argued either orally or in writing to the judge. The disputing parties will not be present as these motions are prepared, as they're responded to, as the judge considers them, or when the attorneys argue them in the judge's chambers. What the disputants get is the fallout—and the bills.

The fallout will come as a ruling from the judge who, operating under his reading of the regulations, or of precedent cases, or of the local rules, or compelled by the seat of his pants, will find for, or deny, the motions of the parties.

Because the rules have been established to give all parties a "fair" hearing, the costs of all the motions, production, discovery and preparation in advance of the trial itself can be very high. Judges will lean in the direction of letting the parties have their say, because the greatest fear any lower court has is to be overturned on appeal for a legal technicality. It is a fear that can raise anxieties about not advancing up the judicial ladder. It jolts judges awake in the wee hours.

Welfare for Attorneys

The litigation involved sums in excess of $150 million. At least 45 subcontractors and suppliers were involved. Every party was represented by a lawyer, aided and abetted by surety lawyers.

An endless number of motions were filed. Sometimes they affected all parties. Usually they affected the disposition of only one matter as it related to one subcontractor.

The Motions Judge established Tuesday morning as the weekly scheduled hearings and/or findings on motions.

The substance of each motion, filed well in advance of the hearings, was apparent on its face. It was typical for only four or five attorneys to have concerns before the court.

Nevertheless, every Tuesday morning the courtroom was packed with attorneys—35, 45, or more—some who flew in that morning and flew out that afternoon—or stayed to play awhile since the court *was* in Las Vegas. The matter took years to resolve.

161

13

This tendency to allow demands for discovery, testimony and document production, and the motions and legal games to oppose them, is why the cost of trial mounts so quickly. Interrogatories will be endless. Everybody will be deposed. Document production will consume a small forest.

Rooms will be filled with attorneys for *everybody's* deposition, or for the judge's findings on motions, even though many of those present will have no questions to ask the deponent and no direct concerns in the motion. All of them will be able to review the transcript anyway. But they all have to eat, and for many of them an insurance company is paying the bills.

If, by the luck of the draw, you and your attorney find yourselves before a fair and competent judge, you will immediately recognize the benefits as he or she puts a short chain on the junkyard dogs. You will immediately sense the judge's directness and common sense, even if you don't understand the legalese behind his/her findings.

If the judge is otherwise motivated or less than effective—prepare for a long siege.

All of these considerations make it critical to have an effective trial attorney on your side, one who is thoroughly conversant with the regulations, the rules of law and the standards that apply to design and construction contract performances. It's also essential for your attorney to be incisive in presenting the applicable law to the court. Quiet incisiveness is usually the most persuasive way to convince a court that specious arguments and junkyard dogs should be put in their proper place. The attorney you want on your side will be distinguishable by this characteristic, which will also be a valuable resource in cross-examination.

But in the final analysis, so much depends upon the competence and disposition of the judge, or the other representatives of the court who will hear the issues in dispute, that your prospects of prevailing may as well be on *Wheel of Fortune* once a claim is formally filed.

Facts v. Obfuscation

Trials operate under a considerable burden of rules. Some of them are helpful to an owner who requires the production of documents he does not have. But not all of these rules necessarily facilitate getting at the truth. Many of them can be used to obfuscate the facts[1]

The frustrating limitation for the layman (non-lawyer) in trial or in arbitration is that he cannot tell a straight story. The facts that are clear and cohesive to you can only be brought out through questions and answers (Q&A). Questions posed by one attorney will be countered by more questions from the opposing attorney. By the time the conflicting Q&As are sorted out, it is often impossible to convey a cohesive continuity of events.

Percipient witnesses (someone who was present at, or was party to the event or conversation) can only testify to "facts." They are not allowed to express opinions or to recount "hearsay" in court. Whether or not the "facts" they recount are *true* facts will be challenged by the other side(s). And since these "facts" can only come out under Q&A, many more hours will go into researching documents to determine which facts are "true facts."

Then, adding insult, injury and frustration, your attorney will tell you that of all the facts you believe to be important, true or otherwise, some are relevant to the legal issues involved, and some are not.

Your efforts to get objective facts heard and understood can become a very disillusioning experience.

The Collapsed CPM "Analysis"

One day you will see some expert's *collapsed as-built CPM schedule delay analysis.*

If you can get someone competent to explain it, you'll find that the distance between *facts* and *truth* will make the Grand Canyon look like a gully. (More about this in Appendix A2.)

[1] *Law, truth, facts: Even a brief scan of Philip K. Howard's* The Death of Common Sense *will confirm that we're all in trouble in the courts.*

13

"Experts," on the other hand, bring the advantage for their client of being allowed to express opinions.

Once you've heard some of the opinions and conclusions some "experts" opine, you will develop a sense of the chasm that can open up between "facts" and "truth."

We will discuss in Parts 8 and 9 the specifications and tools that will help you to respond to these misrepresentations. First though, we'll visit some of the typical alternatives for disputes resolution, and some of the absurdities any of these venues can present.

Obstruction of Discovery

The law is there to get at the truth, right?

But people do foolish things in the course of designing and constructing buildings. They fail to do what they promised (by contract) to do. Or try to transfer the impacts of their failures to others. Or are pressured by others to make assertions there is no possibility they could support with their documentation—or their lack of it.

It's little wonder, therefore, that practiced construction attorneys, abetted by the labyrinth of the law can use an extraordinary number of legal (?) means to delay, confuse, frustrate and outright refuse demands for discovery.

Despite the right of parties to litigation to demand production of documents assured by appropriate laws and regulations, attorneys have been known to hold up delivery of essential documents for years. The demanding parties can only guess at what rites of purification are being performed in the meanwhile.

If an owner has an interest in protecting itself against the unscrupulous, the only way to do that is to establish a continuous, concurrent flow of documentation into his archives throughout both design and construction. And to make effective use of this documentation to create his own trail of Q&A over the development of his project.

Government Agency Hearings

This book is for owners who will at least consider changing the way business is done on their projects. So there's not much here that will be taken under advisement by federal, state or other large government agencies. The immovable mountain will likely remain unmoved. (Although, as taxpayers, we'd all benefit if it could be.)

It's only useful for the general reader to know that many large government agencies have been locked in for years to "binding arbitration." Many federal and state agencies have established claims divisions for hearing disputes (Board of Contracts Appeal).

The warranties and assurances in the various agreements between owner and design consultants and contractors generally apply to government contracts. However, the regulations controlling government disputes in general, and the various boards of contract appeal, are very complex and beyond the scope and immediate concerns of this book.

It is interesting to note, however, that some of these once-frozen protocols have shown signs of moving over into the private sector of construction.

Agencies that have charters to maintain public facilities have experienced fiscal limitations and have had to consider Build/Lease-back Contracts as an alternative to running their own construction contracts. Under these arrangements, private funds are used to build, and to rent back to the government, projects required to serve the public interest (post offices, jails, highways, airports, etc.). These arrangements will, over time, result in some redistribution of the venues and the rules under which construction disputes on these larger government projects will be resolved.

Judicial Arbitration (Jud-Arb)

Just a minute there! You think you're going to trial? Not yet!

Making every effort to unload the congestion of its calendar, the court will, before it even considers setting a trial date, direct the disputants to expend further effort (and money) to "settle" the matters at issue.

In advance of its mandated *settlement conference*, many courts will (in less than Gargantuan cases) require a judicial arbitration. The court will appoint a special master or referee. The special master (SM) (invariably an attorney) is supposed to review carefully the facts in dispute and the applicable law as presented by opposing counsel, and hear arguments.

165

13

The special master's charter is to advise the parties how he believes the matter would be heard by the court if it were to go full term. His efforts are supposed to shed light the parties have not recognized, and to encourage them to settle. Courts, motivated to clear their calendars, tend to put more credance than may be justified in the judgment of some special masters.

Jud-arbs have mixed track records, depending upon the qualities of the special master appointed, as well as the intransigence of the parties. A jud-arb can be a productive exercise. It depends upon how well versed the special master is in the facts of the disputes, the applicable law, and upon how experienced he is in resolving construction realities. It has the same positive and negative potentials as mediation, since the SM's finding is usually not binding upon either the parties or the court.

As often as not a judicial arbitration is an exercise that requires the parties to spend more on legal fees, and pay for the SM's time, but frequently has little more benefit than showing one side some of the legal arguments the other side is likely to make in trial.

If they couldn't talk reasonably before, or through a mediator, the format of a judicial arbitration (which is almost entirely within the discretion of the special master, and therefore subject to the same whims as those of a mediator) is not conducive to changing these dispositions.

But it does keep some attorneys busy without investing in the homework an advocacy representation would require.

Arbitration

From 1950 into the mid-1980s, the construction industry, private and public, was moving in the direction of electing binding-arbitration over trial. For construction disputes, containing as many technical issues as they do, it was believed that the issues would receive more responsive hearing by a panel of experts selected from the industry, that it would be faster, and that it would be less costly.

Binding arbitration requires that the parties accept in advance that they will be legally bound by the decision of the arbitrator(s). The decision will be blessed by a court order formalizing it.

Early experiences established that the arbitration process was usually shorter; required less intense, costly discovery; was less expensive overall; and often produced results more acceptable to the participants than did trial by a jury or court.

On the basis of these results, the American Arbitration Association (AAA) rapidly expanded its number of hearings throughout the U.S. Caseload increased by 482% between 1965 and 1995. The number of AAA panelists increased to 20,000 nationwide.

Panels became loaded with inexperienced, sometimes inept or tired, sleepy-eyed panelists. The AAA has recently cleaned house, down to a roster of 12,000 or fewer panelists, but has meanwhile lost much of its momentum to the initiatives of other venues, notably mediation.

There are distinct differences between arbitration and mediation and other alternative disputes resolution (ADR) venues.

The distinguishing characteristics of arbitration from other ADR arenas are [1] that the arbitrators are usually empowered to make a binding decision; and [2] evidence is presented in a semiformal (or semiinformal) hearing in which testimony is presented, as it is in court, through Q&A.

One distinct downside of arbitration for an owner is that there are no binding regulations that require opposing parties to produce all relevant documents. If comprehensive production of documents, concurrent with construction, has not been specified, the owner can experience considerable resistance, and excessive legal costs, to require the production of documents he believes important to his position. In addition to which, the owner may not even have a clue as to what documents are in the respective files of the other parties. This can put the owner at a considerable disadvantage.

The correspondence between the architect and his consultants, or the reviews of an outside agency returned to the architect's files can often be important to the resolution of a dispute. So can the daily logs and performance records of both the architect and the contractors during construction, or any number of other documents necessary to establish responsibilities for asserted impacts.

The contractor has documents unique to his daily activities. The standard AIA agreements are entirely inadequate as they relate to which of these documents are to be delivered to the owner or to his representatives.

Even if the owner can get the other parties to agree which documents will be shared, the attorneys representing opposing par-

13

ties, or their sureties, can put up all kinds of roadblocks and delays to their production.

The owner can find himself trying to reconstruct history from depositions alone (interrogatories are useless for this purpose). It will require a very effective attorney to build a case from only bare documentation and depositions, unless there has been an obvious breach of the contract or of the law.

Appellate Courts' Reluctance

Despite the inconsistencies, inadequacies, errors of law, arbitrariness and sometime incompetence demonstrated in arbitration, appellate courts have shown a distinct reluctance to overturn, or even to consider, the findings of binding arbitration.

Despite the nominal appearance that arbitration is like a trial (because of the Q&A format), the arbitration hearing itself is not subject to the same rules that a court trial is (supposed to be).

The arbitration panel often makes up its own rules as it goes along—unless these have been definitively established beforehand. And establishing mutually agreeable rules can be a battle itself between opposing attorneys.

Even if he does have many of the documents, the owner is in an entirely different position from a party bringing a claim. The claimant has the initiative. He's had, and organized, most of the documents to support his claim throughout the project.

The owner first has to find out where the claimant is coming from. His attorney needs to determine what tenets of the law apply. The attorney and the owner need to obtain essential documents. Someone has to analyze the facts as they apply to entitlement. They have to analyze how the asserted damages were calculated. Only then can they put together a defense they believe responds to the claimant's assertions of impact.

All of this will require, in addition to the owner's attorney, the assistance of those "other strangers" who were not part of the construction process. They will have to come up to speed from a dead start.

The owner also needs to consider carefully his relationship to the design team and construction manager (if he has one) as they relate to the issues being raised by a contractor. One-time "allies" can quickly become cross-defendants, or even adversaries.

Arbitration usually provides only limited rights of discovery and document production, which is one reason contractors prefer it to litigation. They already have most of the documents they want. They also have many of their witnesses already in harness.

"Hearsay?... Sure."

I had been sitting in the arbitration for two days, waiting to provide my testimony.

As the contractor put on his story through his percipient witnesses, I continued to hear repeatedly what I believed was *hearsay* (what some party who was not present and was not available for testimony was supposed to have said or done). It all went on the record without objection.

While I sat in the witness chair, the panel asked me a direct question. Before I responded, I asked the question in return, *"I can tell you what I heard—would you like some hearsay?"*

The response was, *"Sure."*

The informality of arbitration is usually to a contractor's advantage. Since the owner does not have the formal rulings of a court to control testimony and evidence, it is often much more difficult to get the arbitrators to rule on motions that the owner might consider essential to its defense. Hearsay has been entertained in more than a few arbitrations. Hearsay—*what somebody who can't testify is asserted to have said or done*—can have considerable influence with a panel consisting of a contractor, an architect and even an attorney whose only knowledge of the matter is what they hear. In court, hearsay is disallowed.

Which Forum for Disputes Resolution?

For all the reasons above, plus others, an owner might elect to define a disputes resolution clause for which the venue of choice is a court trial. The clause can contain an absolute requirement for production of documents by all parties, and can contain a provision for non-binding mediation or another form of ADR as a condition-precedent to a court trial.

Such a requirement puts all parties on notice, at the least, that proof of any claim assertions made will be subject to well-defined rules.

13

This, coupled with requirements elsewhere in the administrative specifications regarding the delivery to the owner of comprehensive documentation concurrent with design and construction performance, can provide effective motivation for other parties to negotiate disputes rather than proceed to file a claim.

If the owner elects to define arbitration (or any form of ADR) as the venue of choice, specific production of documents and full access for discovery should be incorporated as a precondition to arbitration.

The owner should never be entirely confident, however, that anything he puts into his contract is conclusive. The courts are notably lenient about giving every party a hearing (paranoia about reversal again). A noisy, aggressive attorney will frequently prevail upon the judge to disregard the very specific requirements in any contract.

An effective judge will restrict the games attorneys can play within the confines of the presumed applicable law. If the judge is not up to this standard, a junkyard-dog attorney can extend a claim indefinitely, even though the facts demonstrate that one party clearly has breached some terms of its performance contract.

An Attorney's an Attorney, Ne C'est Pas?

Many owners—notably public agencies—believe that anyone who passed the bar is competent to represent them in a construction matter. School districts in particular, operating within the agency that incorporates them, put their reliance in county counsel.

That attorney's legal expertise is often limited to personnel disputes, insurance and general risk management. Expertise in real estate law doesn't even tap the nuances of a construction dispute, nor insight into the requirements called for in design and construction contracts. Unless an attorney specializes in construction law he usually has no clue about what's involved in construction disputes.

Even with continuing exposure to construction litigation, what goes on behind those closed doors during design and construction is still a mystery to many lawyers. That's why in this book I've emphasized the importance to an owner of assembling a team that also includes a hands-on construction type—the guide.

Meaningless Interrogatories

There is a process in the law known as *the interrogatories*. Its ostensible purpose is to bring the opposing parties more quickly to irreversible statements of facts that should lead to efficient resolution of the issues.

Uh huh. Ask your attorney (or guide, if your attorney is reluctant) for a set of old interrogatories from a long-gone claim. You will find therein masterful examples of obfuscation, redundancy, filibustering, misdirection—and outright waste of time and clients' money.

In construction matters the interrogatory process generates: [1] a few names and addresses, [2] delays to the overall disputes resolution process, [3] entirely non-productive legal costs, and [4] the destruction of entirely too many trees.

From the standpoint of the participants to a construction claim, however, there is little that can be done. Under the law, interrogatories *must* be answered in kind, chapter and verse, generating more useless legal costs. It's a game that can only be played by lawyers. And it can't be avoided, because it's one of the handy, fee-generating mechanisms that will be employed by any party who wants to delay addressing the central issues.

What If Arbitration is Elective?

Elective arbitration is not always elective. Some contracts stipulate that arbitration can be elected if both parties agree. An owner should not rely on his right to refuse to arbitrate. It is not unusual for a party who wants to arbitrate to prevail with his argument before the court, that arbitration is in everyone's best interests. The court, motivated to lighten its own caseload, may so order and arbitration becomes non-elective.

Arbitration does not have to be under the aegis of the AAA, or any particular sponsor.

13

The disputants can set up their own arbitration rules and procedures with regard to the hearings, the evidentiary proceedings, the discovery and the production of documents. These rules will have to be blessed by the court of jurisdiction, however, so that the results will be legally binding upon the parties.

The Choices: ADR

In addition to, or instead of, trial and arbitration, there is a wide choice of venues in which construction disputes can be heard. The laws, the regulations and the rules that define and govern them are so numerous and varied that discussion of any but a few of them can be both redundant and confusing. They're also subject to change, so the best thing for an owner to do is to obtain the readily available pamphlets distributed by the American Arbitration Association (AAA), by J.A.M.S. (Judicial Arbitration Mediation Services) and similar organizations, and review their implications and requirements with his attorney before negotiating any service agreements.

What is important is that, as an owner, you should know something of the laws that govern available options in the specific state and county (parish or commonwealth) of jurisdiction for your project.

It is also important to obtain some assurance that the attorney who assists in defining the terms of the contracts you plan to use is thoroughly familiar with the rules that prevail in the court that will decide the direction (if not the particular fate) of your project.

It's important to recognize that the *last* jurisdiction in the *last* contract the owner expects to sign (usually the construction contract) needs to be the jurisdiction in the *first* contract the owner signs. There is always the potential that everyone on the project will cross-complain against everyone else. That's why the courts will combine them if you don't.

You, as owner, will want all issues in one venue in any case, since that will enhance your access to documents that may be essential for preparing responses.

When you request your attorney to assist with the agreements and specifications, you will be able to measure his competence in terms of how he anticipates the concerns in this and the two prior chapters.

Mediation—Before and *After* a Claim Is Filed

The proliferation of mediation demonstrates something about our basic human nature.

As many benefits as direct negotiations between arguing parties can deliver, careful, measured negotiation usually requires that a third party create an environment in which the parties will listen quietly to each other. Effectively mediated, they may even come to hear the common sense a neutral party can introduce.

There is a general sense among laymen that mediation is only available once a claim has been filed.

It's an impression encouraged by attorneys, promoted by mediation services and underwritten by the fact that most parties do come together to mediate only after a claim has been filed.

But mediation doesn't inherently require a lawyer to be the mediator, nor that a claim be filed.

In fact, many parties would benefit if, before a claim is filed, they convened with a neutral party, competent in assessing construction disputes, and reviewed their respective positions with a view toward listening to the objective judgment of an independent party. The parties would probably want to resolve a non-disclosure agreement, but that doesn't differ from what any other mediation requires.

This kind of non-binding discussion of the facts, with an objective and competent third party as leveler, can often bring the parties to see their positions more objectively. The principals can remain directly involved, determining the best *business* resolution for their organization, instead of becoming bound up in the legalese that's inevitable (and costly) once one party files a formal construction claim.

This form of mediation can be encouraged by incorporating into the disputes clause of the administrative specifications the stringent documentation requirements suggested above, together with a stipulation that notice be provided to the owner with sufficient time to respond before a claim is to be filed.

13

The potential for pre-claim mediation will disappear once disputes revolving around legal issues, as distinct from negotiable factual issues, are introduced. Once this happens, the controlling rules will become so convoluted that a formal claim is inevitable.

Most disputes are advanced to a claim before the parties are aware that there are other options, because attorneys actively move them in that direction. That's why most mediators are attorneys. They will continue to work to keep it this way.

The Mediation Format

The format of mediation is different from the format for arbitration.

Each of the parties is usually provided an uninterrupted opportunity to recount its position, to present reports and exhibits and expert opinion in a direct manner before the mediator with the opposing party present.

The mediator will then typically separate the opposing parties into isolated caucus rooms. The mediator will then motor back and forth between the parties, extracting their confidential views. He will also, if he's effective, express what you hope are objective judgments of their respective positions to each of the isolated parties.

His purpose—usually expressly stated—will be to get each side to compromise until he senses that they can either be brought to a settlement, or that they cannot.

Construction mediations are typically one to three day affairs, unless they're very large and very complex—in which case they're probably not fit subjects for mediation anyway.

Mediation and Mediators

Mediation, usually of the non-binding variety, can be required by the terms of a contract, or the parties can agree to it, either before or after a formal claim has been filed. The contract may require that the parties mediate before they can proceed to any other venue.

The mediator's capacity to understand the nuances of a complex construction dispute is critical to a reasonable resolution. A competent, effective mediator needs to have these characteristics:

- He needs to be thoroughly conversant with the processes of design and construction and understand the essential interactivities among all participants during construction.

- He needs to be honest and objective.

- He needs to be a quick assimilator of facts and to be able to weed out irrelevant information from the statements of the parties.

- He needs to inspire the kind of confidence that will get the parties to realize their best interest is to resolve matters short of going full term to trial.

- He needs to be thoroughly conversant with applicable construction law; or to have a consulting attorney available to consult who is, if the resolution of the dispute will be driven by legal issues.

- *He needs to be committed to helping the parties resolve their dispute as expeditiously as possible.*

Since this last qualification can sometimes work against the mediator's self-interest, most effective and competent mediators have established a fee structure based on a fixed daily rate or other well-defined service fees that do not relate to hours invested. This type of fee structure can encourage the mediator and participants to resolve differences efficiently because everyone knows in advance where they stand (unlike the open-ended legal costs experienced with many legal firms).

Special Masters

A special master is (almost exclusively) an attorney who has established himself as the intermediary between the disputants and the courts, ostensibly to assist in reducing the court load and facilitating disputes resolution.

The distinction between a special master and a mediator lies essentially in the relationship with the court that appoints, or anoints, him.

There are some fine ones, and some ineffective—or mysteriously motivated—ones, in about the same proportions that apply to arbitrators, mediators and other ADR intermediaries.

Welfare for Surrogates

Those who have set themselves up as mediators, judges-for-hire, special masters, referees or their equivalents are independent

13

contractors. Their livelihood depends upon referrals from the courts and from other attorneys. Referrals depend largely on their success ratio.

The worst of surrogates—and you're not likely to know who they are unless you've been around the campfires for a while—are motivated by:

- Fear of spoiling their track records: They go for the "match the pot" settlement. Identify the maximum one party will pay to settle, and get the other to accept this amount by bearing down on the negatives of his position. Objective assessment of facts, of entitlement, or of asserted damages are not given serious consideration in trying to move the parties.

- Ignorance: They are not committed to the homework essential to resolve specific issues. They disguise with gross legalistic generalities their ignorance of the contract, the codes, the regulations and the standards that should be applied to the issues.

- Mediation incompetence: They have no competence in bringing the parties to candid comprehension of their respective positions.

- Cupidity: They keep things going (as much by inaction as by any exertion) to maintain the flow of income from the parties.

There are some good mediators and special masters, but too few effective ones for our litigious culture. Too many jud-arbs and mediations result in bad resolutions, or no resolutions at all. Too many are protracted because of the incompetence, the non-responsiveness, or the self-interest of the SM or mediator.

The legal process itself is frequently an inhibitor of sensible negotiation and reasonable resolution.

Finding Your Equilibrium within the Legal Process

One more startling blow to your sense of fair play is the dawning realization that trial, arbitration, mediation or any other venue beyond direct and candid negotiation is an *advocacy* proceeding.

THE LAW'S ARENAS: OPTIONS, AGONIES AND ABSURDITIES

The arguments that rear their heads on a construction site are almost friendly compared with the distemper displayed once a formal claim has been filed.

It makes no difference that much of the noise is a put-on. To be effective it has to be consistent, and you won't find many friendly words passing between you and the attorneys representing the opposition.

That means you will hear lies, misrepresentations, exaggerations, and attacks heaped upon your professionalism and personal integrity and on those you retain to represent you.

Attack is the order of the day, and it does no good to respond with emotions.

Because the whole claim process is one of Q&A, it's tough for people who know the facts to get them out cohesively and concisely. The attorneys will bring in "experts" who are prepared to pontificate about the story (fantasy or otherwise) the way it has to be told to support his client's position. It is the unique facility of someone who's been qualified as an expert to share his opinions with all the pontifical erudition he can muster.

But you'll also notice an irony amongst the vitriol. Your attorney and theirs will, as often as not, be on friendly terms. Don't be

Advocacy Environment

The arenas to which claims are submitted for resolution are more conducive to vindication of positions than to reasoned discussions of possibilities.

The Locker Room

It was midafternoon. I thought I was the only one in the locker room. Then I heard two voices speaking low on the other side of a row of lockers. In a few words, it became obvious they were attorneys on opposite sides of a case.

"I could bill a few more hours on this thing if you could."

"Yeah, well my office isn't so busy right now either, and I think I could stretch it out to the end of the month. I know the insurer's good for it."

"Okay. Why don't we schedule a negotiation with my clients in about four weeks and bring them to settlement? I already know I can get them down to what your company's willing to pay."

"Let's say you call my office to set up a meeting for the 25th?"

"That day's clear on my calendar. I'll do that."

177

13

surprised. They frequently meet, and often cooperate, on other occasions, other legal matters, in other venues.

It takes a while to find your center within the mini-storms, disdain, contradictions and inconsistencies that erupt in this advocacy environment. It can help considerably to have on your side an attorney who will share patiently with you, in terms you can understand, the developing positions of the respective parties.

Justice, Fairness and Expectations

It's important for an owner and anyone else who steps into a legal arena—including a court trial—to realize that the outcome of a construction claim will be decided by an indeterminable amalgam of competence, honest effort, fear, laziness, capacity or incapacity, and self-interest on the part of many of the participants.

The rulings of the court (the judge) will have a direct effect on how the facts are presented, admitted, dismissed or tried. Rulings, for example, to admit evidence, to recognize the potential for summary judgment, to direct the production of documents, to admit or deny jury instructions.

A judge's greatest fear is to be reversed.

If the judge is up to speed on which regulations apply to specific motions and is willing to apply the correct ones despite the bared teeth of one of those junkyard dogs, the party who has some truth to his argument may have some chance to prevail.

But some courts lack the backbone to face down heavy confrontation. A judge's greatest fear is to be reversed.

So the surrogates the courts appoint reflect this same fear. Special masters, acting as interim hearers (as distinct from triers) of facts on behalf of the court, are notable for their reluctance to make any legal finding that will conclusively resolve a matter.

Both because it might become the basis for reversal by the court that appointed them. And because it might abruptly end the matter, cutting off a source of income to the special master.

As a result, one of the purposes for which courts established the special master system, to reduce the burden of the courts, often frustrates the basic purpose of the law: to attempt to do justice.

You, as one of the parties, aren't likely to know which vice—laziness, cupidity, fear or unwillingness to confront that junkyard dog—negatively impacted your case. The special master's, and potentially the court's rulings, will be cloaked in legalese only your attorney can challenge.

Effective Negotiation

If the owner has had effective representation throughout the construction process, and has acquired comprehensive documentation of the respective contract performances on the project, the least costly and potentially most productive way to resolve a construction dispute is through direct negotiation with the other party or parties.

Many owners who have a construction dispute to settle approach it like any argument they might have with a family member. This does not often produce useful results.

Effective negotiation means much more than just sitting down to argue your case with the other party.

Even if you believe you have right on your side, it pays, before trying to negotiate a settlement with someone who's at arm's length from you (the opposite party to a contract), to try to see the matter from their side. Effective negotiation starts with some serious listening. Listening without predisposition or retorts to the other guy's argument.

A lot of people think that listening means agreeing. That's shortsighted. It means that you're giving the other party the opportunity to get everything off his chest. That can relieve a lot of tension.

In order to listen well, it's important for you to know, first, how good or bad your own case is.

Then you can start talking from a middle ground from which, without necessarily agreeing to anything you hear, you can at least start a dialogue. Get to that point, concede at least some part of what the other guy asserts and you're already halfway to a solution. Pollyanna didn't just pop up here. Every effective negotiator will tell you the same thing.

In the matter of a construction dispute, it means that you start, not the way you might with an emotional or personal argument, but by undertaking an objective evaluation of the strengths and weaknesses of your own position, and the same evaluation of the other party's position.

Many owners who have a construction dispute to settle approach it like any argument they might have with a family member. This does not often produce useful results.

Effective negotiation starts with some serious listening.

13

With that in place, you can start listening seriously, and searching for a middle ground from which to work.

Because of the potential complexity of a construction dispute— and particularly because emotions and egos can interfere—it is usually productive for the disputants to find a good *informal mediator:* one through whom the parties can communicate and who can help them find that middle ground.

Five more points about effective construction dispute negotiation:

- The more homework you do, the closer you think you are to being ready to go to court, the more effective your negotiation will be, because you'll know more about the details than anyone else.

- Identify a single spokesperson who has a thorough handle on both sides of the issues. Multiple representatives dissipate the energy and the cohesiveness of a party's negotiating positions.

- Your attorney isn't always the best spokesperson in negotiations. Attorneys tend to emphasize the rules of law and pre-emptive discussion of dollars, instead of the facts that support or defend against entitlement arguments. Attorneys are sometimes best employed as a legal resource in a negotiating environment, much as any other expert.

- Always have a couple of points you know you can give away, even though reluctantly. It will create a credit balance on your side of the ledger as you move down through other issues.

- Set up in advance at least two intermediate money points and a final no-further point before you start. Use them to help push the discussions past difficult roadblocks.

The following story is representative of scores of projects and claims on which clients have unknowingly been represented by attorneys whose conceit has exceeded their grasps of reality. The frequency with which this conceit shows itself in all phases of construction projects, as well as in negotiations of claims, should give any owner (contractor, or any other party) pause about how he or she will be represented once the concerns of his project pass into the unchallenged control of a lobotomized lawyer.

The Lobotomy at Work

It was a meeting to plan negotiations with the state agency to recover damages on a $40 million lump-sum contract for a state medical facility. The contractors had brought a $9 million claim for recovery on the basis of design defects, administrative failures and the consequent delays and extra costs.

An attorney who had been selected by one of the principals' attorneys for his "excellent track record dealing with the state government" was directing the meeting attended by experts, contractors, attorneys and principals.

The "lead attorney" proposed to compromise on the $9 million from the outset of the negotiations, and ask for $5 million. (This starting point would clearly help to preserve his "track record.")

At first no one demurred. The "leader" had made it clear who was going to pontificate at both this meeting and tomorrow's negotiation.

The "leader" asserted that he believed that "the delay issues were the result of the inefficiencies imposed by the design," and produced a Mechanical Contractors' (MCA) manual. With reference neither to the facts, nor to actual conditions on the project, the "leader" selected the inefficiency factors in the MCA tables that he asserted would best achieve the amounts he proposed for the negotiations. He then proceeded to add up MCA factors that were inherently non-additive, and accounted for inefficiencies of 41%

No one spoke, but blank looks passed around the conference table as the "leader" continued to apply factors altogether inappropriate to the conditions on the project, and compounded his ignorance by applying mechanical inefficiency factors to the work of other trades.

Then, as the "leader" took a breath, one consultant took the bit in his teeth and asked a couple questions: *"How and when did $4 million disappear from the $9 original million claim? Are these the factors we should be applying? The numbers we've added aren't additive."*

In a tone that broached no further questions, the response was: *"I'm the attorney here. You're not here to ask questions. You're here to answer questions if I need assistance with the state people."*

Fortunately one of the principals got the wake-up call. When the team went into negotiations the next day someone else led the presentation, started from the bases asserted in the original claim, and effected a reasonable resolution of the issues with the agency.

13

One Absurdity—Two Realities

After a year of wrangling about the construction of a science building, the contractor and university met for a hearing in front of a Disputes Review Board (DRB). The DRB consisted of three men who had attained their roles through seniority with various public agencies.

The university was represented by staff counsel, outside counsel, its project manager and two claims consultants. The contractor was represented by its principal, two attorneys, and a construction advisor. An independent consultant was there as an observer.

The central issue boiled down to whether the contractor should be excused or compensated for the delays resulting from changes required in the structural steel design.

The observer advised the Board that there were three possibilities:

- The design changes were within the control of the contractor. The delays were therefore the contractor's responsibility. (Unlikely, based on the facts presented).

- The changes were not within the control of the contractor or the owner. The contractor was, at the least, *excused from responsibility for the delays*.

- The design changes were within the control of the university. The contractor was both *excused* and should be *compensated* for the delays.

The university's counsel objected: *"There's a fourth possibility. The contractor should have re-sequenced his Work to compensate for the delays."*

Absurd as this assertion was, the DRB appeared to give this position serious consideration. The parties reached no settlement and ultimately went to trial, where the contractor prevailed.

Epilogue: This experience demonstrates two realities. [1] More than a few panels are neither aware of applicable construction contract law, nor competent evaluators of construction realities. [2] When an owner awards a lump sum contract without limitations on the contractor's sequencing of the Work, the owner surrenders control over how the contractor plans his Work, as long as he complies with the terms of the contract. The owner cannot, in correcting design errors, require a contractor to re-sequence his work without compensating the contractor for reasonable costs.

6

TIME: THE IRREPLACEABLE RESOURCE

If the forces waiting to destroy this irreplaceable resource are to be controlled, the owner first has to make the irreplaceable resource visible.

14

IT'S INVISIBLE—SO EVERYONE SEES IT DIFFERENTLY

Time Is of the Essence

All design and construction contracts should stipulate that *time is of the essence of this contract*[1].

Black's Law Dictionary says that this means that *"a failure to do what is required by the time specified is a breach of the contract."*

It's a matter of contractual responsibility for the owner, the design consultants and the contractor to have a measuring device for **time** and for each of them to **ensure that he and his representatives adhere to the reasonable schedules that are warranted by the construction documents.**

That's precisely why it's such a wonder that contractors use people with limited perceptions of the construction process to prepare their CPM schedules.

It's a reflection on the management capacities of design consultants that few are competent in scheduling theirs or anyone else's efforts.

And ironic in terms of the exposures it presents, that owners do not make an investment in learning what time, its misuse, its productive measurement, and the reporting of it are all about.

Only a minority of those professionals continuously involved in construction have troubled to learn anything about the benefits of using the powerful graphics tool known as the *critical path method* to refine the planning of their responsibilities.

It's another wonderment that so many owners and contractors accept the things that "schedulers" produce to "comply" with the contract specifications.

And it's a major irony and concern that the folks who are paid extraordinary sums to *advocate, arbitrate* or *mediate* time-related construction disputes invest so little concern in knowing how to read and challenge the junk that's presented in the guise of *"as-built critical path analyses."*

[1] *But see the AIA commentary recounted on page 261.*

185

14

Time: **The Essence is Different for Everyone**

The three most useless things for an airplane pilot are:

- Altitude *above* him,
- Runway *behind* him,
- Ten seconds ago.

That's why a pilot prepares a flight plan *before* he leaves the ground. Time and space already spent can't be recovered.

As the airplane analogy suggests, everybody "sees" time in a different way.

If an engine quits on takeoff, the pilot has ten seconds to do the right things.

You measure your own time in terms of meetings, birthdays, weekends, vacations, bond issues and payments due. You plan in terms of your watch and a monthly calendar. You expect to occupy your building project 19 months after construction starts.

"They"—those strangers—are adding up their time in hours, multiplying by hourly rates, projecting paychecks or billings. They plan in terms of *working days*. As they expend those hours and days they're using your money:

A crew of 6 workmen has an average rate of $28.46 per hour. The cost of eight labor hours for this crew equals 8 x 6 x $28.46 or $1,366.08.

A delay of 4 days equals 4 x $1,366.08, or $5,464.32.

The contractor's markup (25%) brings this to $6,830.40.

The delay by your architect in responding to an RFI on critical work just cost you **$6,830.40.**

If your project experiences eight similar delays over the course of construction, your contractor can justifiably claim an additional **$54,643.20** against your project. No value will be added.

Alternatively—

Just one four-day delay that holds up all the labor on the project—a total of 34 people—will cost 34 x 8 x $28.36 x 4 days x 25% markup: **$38,705.60.**

Eight similar delays over the course of your project could bring the total delay costs to **$309,645!**

Those delays on your project were moving along behind those squares on your monthly calendar. But you didn't see them because they were invisible until they became nonrecoverable.

Making Time Visible

All manner of things are attached to this invisible resource. The carrying costs of the money you will spend. The loss of income associated with a delay. Your own extended administrative costs. The contractor's field-office and home-office overhead costs. Equipment costs. Possible escalation of labor and materials costs related to delays. And the costs that gargoyles and the disputes-resolution industry will bring into your nightmares if *time* becomes a central issue of disputes.

There's one essential truth about *time:*

> *If you're going to control how it's used, you've got to make it visible.*

Like Indiana Jones casting sand on that invisible bridge in *The Last Crusade.*

If you allow time to remain invisible, it—and your money—can be misspent behind those closed doors we've talked about.

What you need is:

- Enough sand to cast out there on the invisible bridge of time on a regular, continuing basis;
- To know what to do when you see how others are misspending *their* time and *your* money.

Management of Time
Requires Measurement of Time

Making time visible is one essential for managing it. Being able to measure how others use it is the other essential.

> If you're about to sail an ocean, when do you start to measure where you are? Only after you've been at sea a couple of days? If your first port of call is 8 days away and you wait until the 2nd day to check your location you could be on your way to Newfoundland instead of the Azores.

When your architect asserts that the "Schematics are 75%" or the "Working Drawings are 50%" and asks for payment, it's already

14

The Costs of Time Lost

- The projected construction cost of your project is $4 million.

- If you add up the architect's fee, land cost, construction cost and management fees; plus construction loan costs for a year; and then multiply by 10% and divide by 365, you'll find the cost of money that will go into your project could be $2,000 per day. This carrying cost (i.e., interest costs) of the money that will go into your project is real money even though you don't see it added somewhere in a column on your budget.

- Now, let's add up delays. Your architect completes the construction documents two weeks late. He adds another two weeks to the bid period to allow for corrective addenda. Then your "committee" changes its collective mind (oxymoron?) and other addenda are issued. The bid date is almost two months later than you planned. This pushes the project into possible winter conditions (but we're only considering money costs, here, not the other "impact" costs).

- The contractor is delayed a couple of months because of additional changes required during construction.

- The completion of your project is delayed 4 months beyond what you planned, all attributable to design changes.

- If your cost of money invested is $2,000 per day, and your project is delayed four months, you have incurred a "hidden" expenditure of $240,000. There is no offsetting value to your project. And this is in addition to any direct costs of changes, delay costs or loss of potential income.

Then Add This:

- The contractor has claimed 20 days extension (part of his two months of delay) as result of the design changes. The changes cost $23,000. The architect recommended approval of the change orders. You signed it. But neither of you negotiated the rate for the 20 days.

- At the end of the job, the contractor presents you a bill for 20 days times $1800 per day for field and home office extended overhead costs. That's $36,000 added to the $240,000, with no additional benefit to your project. It does not cover the $23,000 of the changes themselves.

- And we still haven't accounted for the impact costs of starting the job on the edge of winter instead of during the summer.

too late. What will you know about their coordination, the correlation of the technical specifications, or their compliance with your budget? What is "50%"? You will search in vain in the AIA design agreement for a definition.

The plumbing subcontract is $224,600 and the electrical is $238,950. When the contractor reports that the "Plumbing is 65%" or the "Electrical 40%" in a plant or hospital addition of 150,000 square feet, what will you be paying for?

There is a tool that can make the measurement of performance during both design and construction visible. The *real* critical path scheduling method.

Making Interactions Visible

Recall our discussion about *interactivity* in chapter 4? That the owner is usually occupied with the visible interactivity of payments and changes, but needs to become more aware of the invisible ones that occur between consultants during design, and among everyone on the project during construction?

The measurement of time on a construction needs to be defined specifically in terms of those interactions. To see and measure those interactions the owner needs to have an outline in the sand that defines any time that *work, or questions and answers, will be handed off from one person (or firm) to another.*

Whatever system the owner uses to cast the sand in which to draw the originally scheduled interactions, that system also needs to provide the facility to report what actually happens to those interactions.

And since there are a lot of people interacting continuously in different locations, on different parts of the project over its duration, the system needs to provide a readily assimilable picture of what's happening.

The system needs to be able to present information *graphically.*

The need for this kind of system for scheduling projects was recognized a long while ago by a man named Gantt. He invented the bar chart in the late 19th century. Bar charts are typically plotted on an "X," "Y" grid. "X" runs along the horizontal scale; "Y" runs up the left side vertical scale.

14

They follow the format of Figure 14.1, in which work tasks ("activities") are listed down the left side ("Y") of a chart and durations along the horizontal ("X") scale. The increments can be hours, minutes, days, weeks, years or whatever best accommodates the work being scheduled. Days, weeks and months are most common in construction.

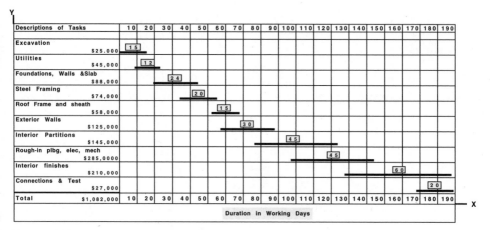

Figure 14.1

In Fig. 14.1 the durations of the activities are shown at the top of the lines that represent the scheduled durations of the activities. This chart also includes the dollar values budgeted or estimated for each activity.

It is important to be specific when days are used in bar charts. Figure 14.1 shows durations in working days (WD), considered to be based on a 5-day week. Using an approximation that the number of calendar days equals the number of working days times 7 and divided by 5 (i.e.,7/5), the first activity in 14.1, "excavation" (15 WD) covers a span of 21 calendar days.

The bar chart, despite its several limitations, served the construction industry into the late 1950s.

PERT and CPM

Then in about 1957-58, the predecessor of the critical path method (CPM) was invented. It was called PERT.

A brief history of PERT and CPM is included in Appendix A-2, which is intended to give you a very synopsized introduction to how to read and use this powerful tool for planning and measuring contract performances.

With PERT/CPM the industry was, for the first time, provided with a tool that could be used to develop, analyze, report and measure the separate actions of individuals, crews and firms as they needed to interact, and actually did interact, when designing, estimating, building and documenting construction projects.

The industry immediately took to the concept (originally developed in the military and in industry, as so many other systems assumed into construction have been). For about 10 years (1960-1970), it was applied vigorously and effectively to the planning, scheduling and reporting of construction activities.

In the past 25 years CPM has had a checkered history. But you can read more about that in chapter 16 and in Appendix A-2.

What we're concerned about here is what *can* be done with the effective use of CPM.

Making Visible What's Invisible in the Bar Chart

The bar chart in Figure 14.1 presents *conclusions,* but not the thought processes that arrived at them.

In order to develop 14.1, someone needed to think about the resources, labor, materials, logistics and resultant costs that would produce these relationships and durations. In reading this bar chart, you receive none of this information. As result, you can't challenge its reasonableness, propose improvements, or propose an adequate basis with which to control either payments or costs.

Additionally, the components and durations of some activities are clearly extensive and interrelated (there's that *interactivity* concept again) but the bar chart provides insufficient information to evaluate whether *interior partitions, interior rough-in* or *interior finishes* can be performed in anything like the time and costs shown for them.

191

14

It's also apparent that some tasks begin at the midpoints of the tasks that precede them, but these points aren't identified.

It would be very helpful to break these tasks and costs (and potentially the resources, logistics, support equipment, deliveries and administrative support required) down into bite-sized pieces, to see the relationships of the pieces, and to so manage the project that the pieces will come together in timely, efficient ways.

That's what CPM has made it possible to define, to show graphically, and thereby to analyze in detail.

A first cut toward breaking these activities down into detail is shown in Figure 14.2. In this chart, the time by which a task *lags* behind the start of the activity that precedes it has been shown explicitly on the chart.

This *lag* is also called a *lead time*. Seen from the viewpoint of the follow-on activity, it is the time by which the preceding activity *leads* the following one. In chart 14.2 this does not specifically tell us *why* this lead time is there, but once it's defined you begin to realize that each bar is made up of other relationships that are (temporarily) hidden in each bar.

The *exterior walls*, for instance, include, but do not yet show *formwork, reinforcement, imbedded items, complete formwork, place concrete, strip forms.* Also, since the 30 *working day* duration is equivalent to 45 calendar days, it would be useful to break the work for the walls down to show how much will be done in the first month, and how much in the successive month, in order to measure progress and payment.

As Figure 14.3 demonstrates, this start into breaking down activities also demonstrates a beginning definition for the *controlling work. It has now become evident, by defining the lead times that only the first parts of the bars control the overall duration of the project: the parts that flow continuously to the end of the project.*

The second parts of the bars are *not* controlling, in that they don't need to be completed in order for the work to continue to the end of the project (as we'll see in a detailed CPM analysis, this is only true within limits, of course).

IT'S INVISIBLE—SO EVERYONE SEES IT DIFFERENTLY

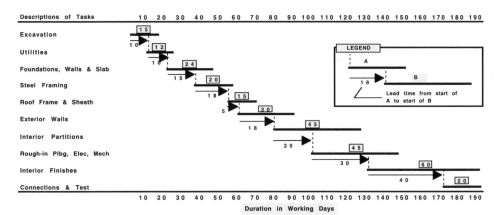

Figure 14.2

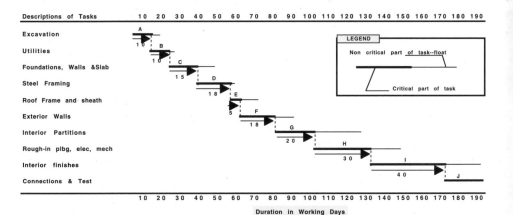

Figure 14.3

14

As we'll see in Appendix A-2, CPM will continue, through the process we've started, to make this detail available *if* the basic rule applied is that *CPM activities will be defined so that they represent each break when the work is passed from one trade or crew to another.*

Using this rule, the original bar chart in Figures 14.1 and 14.2 will become CPM analyses with 40, 60 or more tasks.

Clearly, we need a better tool than a bar chart to handle this amount of detail. That tool is CPM and in Appendix A-2 you will see how its application to this kind of project will develop quickly into an effective tool for the architect's and contractors' planning.

Very quickly, then, the bar chart that originally had only 10 tasks can become a chart with the detail of 40, 60 or more tasks. At that level of detail, the chart starts to provide more useful information, but it also becomes unwieldy to develop, to read, or to use as a tool for planning or measurement.

It is at this point in the advance of the CPM technique that the computer comes to the rescue. With a computer program available to accumulate, calculate and sort out the extended levels of detailed relationships, it becomes possible to develop a graphic chart with as many relationships as are required to plan, schedule, measure and manage a project effectively.

It is entirely possible to put hundreds or even thousands of separate tasks into a CPM schedule and to use the computer to calculate their projected performance dates. Then, the computer can be used to select, sort and print work activities in useful lists, selected by trades and areas, and to print these relationships in easily read color-coded, time-scaled charts.

In fact, the computer makes it possible to print even complex *CPM schedules* with thousands of activities in time-scale printouts, without showing the connections in the original CPM network analysis itself.

As long as the logical relationships and durations in the *network analysis* are carefully maintained, the bar chart the analysis produces doesn't need to show those relationships. All that's needed is to establish and maintain the credibility of the diagram as backup for the computer printout.

194

The computer can calculate what it is fed in terms of logical relationships and print the results in time-scaled graphs, which can be read exactly as a bar chart is read.

This eliminates one of the major objections to the maintenance of a CPM schedule: the need to maintain it in time scale—a tedious and unnecessarily time-consuming effort.

A Tool for Analysis and Planning

If the person using the CPM technique understands the processes being planned and scheduled *from the viewpoint of the participants who will be interacting,* CPM can be a powerful analytical tool.

The relationships of these interactions can be drawn, durations can be assigned to each task, and aggregate durations can be calculated between important points of measurement ("milestones"). These aggregate durations of grouped tasks can be tested against experience and the availability of resources. They can be changed and tested again until the optimum relationships and durations demonstrate a reasonable or optimum way to proceed with a project or group of tasks.

Once the optimum relationships and durations are projected, the resources required can be associated with respective *activities.* Each activity can then be assigned a cost. Codes can be associated with respective activities that identify *who* is performing the work, and *where.*

All of this data can be input to a computer (one that has an appropriate software program), to produce time-scaled graphics. Assignments can be selected and sorted so that information can be distributed to those who need to perform those selected tasks.

Since it is possible to associate costs, or allocation of payments, or use of resources with the activities in the data, the computer can also produce cashflow charts, payment schedules and resource-allocation reports.

Figure 14.4 is one example of the type of an analysis that can be undertaken. This chart demonstrates one of the exploitations that can slip by if this type of analysis is not done. In this case, the owner projected a reasonable allocation of payments prior to bid. The contractor provided his own allocation for payment. The analysis demonstrates that, if it's accepted as presented, only 10% of the base contract amount will be available to complete the last third of the project duration. The general contractor will have gotten all his costs and profits up front, and will have little motivation to expedite the performance of the subs at completion (which anyone who has ever built a project before will recognize as a classic concern).

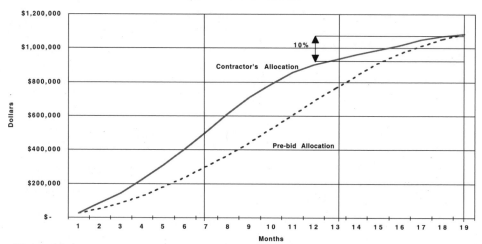

Figure 14.4

Who Does *Time* Belong To?

It's absurd to treat *time* as though it were a doll that can be tossed about, discussed at arm's length, and torn apart with each contracting party on the project arguing about how much is theirs.

Time isn't made up of pieces and parts—it's all of a piece. If the participants try to break it apart, to view it as it would be seen in the shattered pieces of a single mirror, all that remains is arguments waiting to happen.

The success of a project depends upon keeping that view together.

Time on a construction project is everyone's resource. The only way to get a good picture of it and to set up an effective way to measure it is to establish at the start of the project a cooperative environment and an effective system in which the use of time by *all* participants is established for the rest of the project.

A Tool for Measurement

We have now made the *planning and the scheduling* of time assignments visible.

We need to make the actual performances of those assignments equally visible.

For this we need more than the CPM schedule, but the parameters and the coding we defined in the CPM will establish the details and the levels of the information we need.

We have coded the CPM work activities so they can be selected by person, by crew or firm. When important, we have also encoded the location of the work. Money (outflow or inflow) has been similarly defined.

To measure performances, then, we need to receive information which is reported with these same parameters.

If we have scheduled the interactivity of design work so that it can be measured at intermediate milestones, it is important to be able to measure how these interactions are progressing. That's the purpose of the Buildability/Budget Analysis (B/BA) milestones we referred to earlier, and which we'll expand upon in Part 9.

Under the construction contract, we have required the contractor to integrate the scheduling of shop drawings and submittals into the construction schedule. We need to receive on a regular schedule from both the contractor and the architect the logs of how these submittals, plus RFIs, ASIs and changes, are being processed.

If there are changes or unforeseen events, we need to receive from these same participants the records of when and how these events are being handled. These reports will include logs of transmittals and reviews of submittals, as well as records of events and problems as they're reported and discussed in project meetings.

The contractor's daily reports, and the inspector's reports need to incorporate the number of bodies and the amount of equipment actually used on the project (which is all many daily reports show). But they also need to report the number of crews, by trade, where they're working, and the quantities of work and materials installed. Plus unusual events, problems and downtime.

Time on a construction project is everyone's resource. The only way to get a good picture of it and to set up an effective way to measure it is to establish at the start of the project a cooperative environment and an effective system in which the use of time by all participants is established for the rest of the project.

197

14

None of this amounts to extraordinary reporting, though some contractors will complain that it is. It's a basic requirement on carefully managed and documented projects.

Once this kind of data is delivered, the owner may take either of two actions:

1. If there is no urgent need to evaluate events or to take timely action (which should become apparent from competently maintained weekly reports and meeting records), the owner may just organize and archive it.

2. If the owner needs to take some immediate action, it may become necessary to organize this data in a computer database, and even to start the development of an as-built time record of the project. (This is another time when the availability of the owner's guide is likely to be important.)

A computer database is a very simple tool.

All that's required to organize and access the data required to document a construction project is a *flat-file* computer database. Anyone competent in the use of the scheduling software described above should be able to set up a useful database. Effective scheduling systems have this type of database integrated into their software.

The purpose of the database is to organize the data relevant to contract performances on the project, and to provide selective access to this information in response to events that require action. The coding of data suggested above for the CPM scheduling information will provide a basis for access-coding of this data in the database.

One use that can be made of this information is shown in Figure 14.5. Here two contractors' uses of labor hours have been charted against their overall project durations. The labor curve on an efficient project generally assumes the shape of a bell curve. The labor curve on an inefficient project often shows that the contractor does not use labor efficiently at the beginning of the project, and then needs to accelerate (usually incurring losses of efficiency) at the end of the project. Inefficient management of construction often results in delays, mistakes and poor workmanship. Early evidence of this inefficient usage of labor is usually a warning sign that the contractor will search for ways to put the burden on the owner.

Comparison of Project Efficiencies on Basis of Labor Hours Utilization

Mar Apr May Jun Jul   Aug   Sep   Oct   Nov   Dec   Jan   Feb   Mar   Apr

Total Hrs

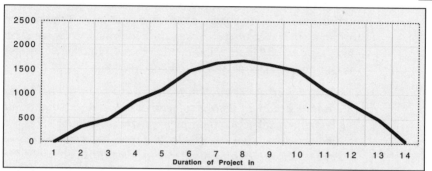

Duration of Project in

| Efficient | 0 | 320 | 480 | 860 | 1080 | 1480 | 1640 | 1700 | 1610 | 1490 | 1110 | 800 | 500 | 30 | 13100 |

Labor Hours Expended Per Month

AN EFFICIENT PROJECT

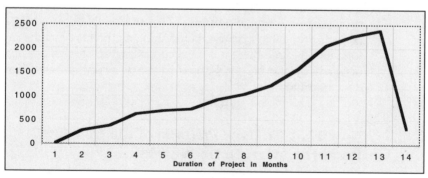

Duration of Project in Months

| Inefficient | 20 | 285 | 385 | 640 | 700 | 740 | 940 | 1050 | 1240 | 1590 | 2060 | 2260 | 2380 | 340 | 14630 |

Labor Hours Expended Per Month

AN INEFFICIENT PROJECT

Figure 14-5

199

14

Project Reconstruction

It is sometimes useful—and in response to time-based construction disputes, it is essential—to cast the historical information from these various reports into a graphic format that parallels the time-scaled graphics of the original schedule.

It is very helpful, for instance, to be able to compare events as plotted in time scale from the project schedule, with the actual performance of these events in the same time scale. This amounts to putting a bar chart with *scheduling* information above a bar chart with *actual* information relating to the same activities, and measuring correlations and differences (see Figure 14.6).

> *The project reconstruction that results from organizing actual or historical data is sometimes called an as-built CPM, but this is an inaccurate concept. By definition, CPM is a planning tool that necessarily incorporates some discretionary relationships and some elective durations, controlled by available resources or by commitments made by multiple parties (consultants, subcontractors, suppliers, etc.).*

Organization of actual data into a time-scaled chart is just that: historical information. As such, options have been eliminated. It is simply what happened. It is better called a *project reconstruction*.

Accurately transcribed from source data, a project reconstruction should be an irrefutable picture of what actually occurred on a project. Irrefutable because, before any interpretation of the information is attempted, it is an accurate picture of what actually occurred on the project (subject to the accuracy of the source documentation).

Now That We Have *Pictures* of the Invisible Resource, What To Do With Them?

Imagine that we have:

1. A competently developed, adequately detailed CPM network analysis, or schedule, that has been produced in the form of intelligible lists and charts.

2. A competent and comprehensive accumulation of historical data, supported by documents organized within an efficiently coded database and charted in the form of an as-built (project reconstruction) chart in time scale.

IT'S INVISIBLE—SO EVERYONE SEES IT DIFFERENTLY

A PROJECT RECONSTRUCTION
How Documentation Can Be Correlated To
Demonstrate Relationships of Causes to Events & Impacts

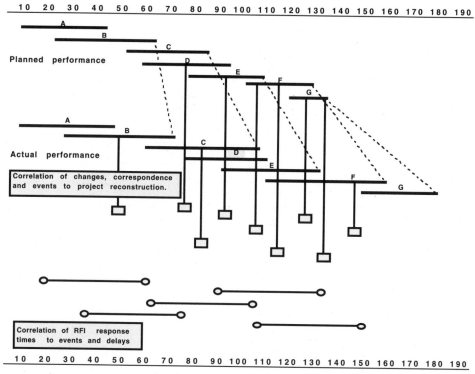

Figure 14.6

201

14

3. A series of time-scaled charts competently developed[2] from this data, so that the historical performance of the duties and responsibilities under respective contracts can be compared with the commitments made in the original schedule, or to updated schedules that have been accepted by the owner.

These historical as-built charts can contain correspondence, dates of transmittals, events related to changes, actual performance of work, or status of submittals, problems and unforeseen events as reported by project documentation.

Organized so that the events in these charts can be compared with contract requirements and commitments in the planning chart or charts, it becomes a matter of insight and judgment to diagnose why and how certain disputed events have occurred.

Undertaken conscientiously, and without a biased attempt to prove some predetermined conclusion, the efforts described above can be a powerful means with which to resolve disputed issues.

What Have We Accomplished?

More than is immediately apparent.

We have, of course, found an effective way to make time visible.

Additionally:

- By incorporating the requirements for scheduling both design and construction into a system that makes time visible, we have acquired a means to ask meaningful questions of the contracting parties (architect and contractor) who will be required to produce these schedules.

- We have acquired a tool with which to *Program* and *Plan* our own project even before we hire those design and construction teams. By using these tools *in advance* of committing to those contacts, we have developed insights about what we expect from those contacting par-

---

[2] *"Competently developed" is a very meaningful concept here. It does not mean that as much detail as possible has been incorporated into these studies. It means that experience and intelligence have been applied so that the resulting organization of data is demonstrative of useful information that can be intelligently applied to the matters at issue.*

ties *before* we contract with them. This makes it possible to define and resolve more meaningful Requests for Qualifications, Requests for Proposals and Agreements. We have moved well beyond the simplistically flawed *mission statement.*

- By using these tools concurrently with design, we can accomplish the central purpose of this book: to *build the project as it's being designed,* in the form of a pre-construction CPM analysis of construction logistics. This, in conjunction with a detailed cost estimate, can be used to test the buildability and Program compliance of the construction documents *before* they're issued.

- In conjunction with the homework we will (once we're through Part 9 of this book) have invested in reading drawings ("bluelines") and specifications, we will be equipped to ask intelligent questions at each of the B/BA design milestones we've defined in the architect's agreement. In the process, we should have eliminated or substantially reduced the 80% of construction problems caused by defective construction documents, and prepared ourselves to respond to the 20% that couldn't be foreseen.

- We will have developed definitive perspectives, a detailed schedule and a pre-bid budget at levels of detail sufficient to test the responsiveness of bids we receive before we commit to the construction contract.

- We have found a way to share realistic perspectives of *time* with everyone on the project. We should be able to avoid arguments about who it belongs to, and the typical owner indecisiveness and mind-changing that plagues so many projects.

- And throughout the construction process we will be positioned to ask intelligent, challenging questions of the design consultants and contractors, and to provide our own timely responses when they're required.

In short, we will have become better informed about the overall direction of our project than will be some of the people we retain to design and build it.

And we have started a process that can have everyone *seeing* that invisible resource in ways that are consistent with the owner's goals for project success.

14

Your Building Needs A Bridge

By the time you've executed the construction contract your project is already fading from the concerns of your design consultants. They have their businesses to maintain. The *provocation* of other clients is creating a certain amount of *chaos* for their collective *muses*. The chaos that may be accumulating out on your project has lower priority than the drawings currently on their drafting tables.

Unless you have constructed a bridge between them and your contractor, and unless you have in place an active maintenance and traffic-control program, a canyon will open between your design consultants and your project that could swallow your budget.

The name of your bridge is *Active Continuity*. The active continuity of *measurement* and *documentation,* together with the assistance of someone you trust to keep you informed of the actions and decisions you need to make.

15

DESTROYING THE IRREPLACEABLE RESOURCE

Delay/Disruption Claim

The Delay/Disruption Claim

The most effective destroyer of a project's resources of money, time and productivity is the delay/disruption claim.

The Recurring Nightmare

Most of us have had the nightmare that we're being chased.

Some of the people chasing us are *almost* familiar, but most aren't. Some of them are telling you which way to go, but every time you turn you keep running into more strange faces. They're all coming your way, shouting unintelligible words. You can't ever seem to run fast enough to get away.

Imagine that nightmare every night for months, waking every morning in a sweat, and every few nights you're presented with a bill you have to pay just so you can keep running.

Welcome to the nightmare of the construction-delay claim gone full term to trial or arbitration.

How It All Started

Construction is a freight train. An inexorable force, like gravity. An avalanche crashing down a mountain. Once set in motion, there's no stopping construction time.

The expenditure of construction time brings with it increasing exposure to the potentials of conflicts. Conflicts in the drawings. Conflicts between the designers and the contractor. Conflicts between the contractor and his subs and suppliers. Conflicts in the perceptions and responses of the architect, the contractor and the construction manager.

Conflicts the owner may not even know about while they're happening. He usually doesn't, or else hears about them late.

A delay claim asserts that the construction-contract duration has been extended as result of conditions beyond the contractor's control.

A disruption claim asserts that the contractor has been unable, due to conditions beyond his control, to perform contract work as efficiently as he originally planned and bid it.

These claims are often associated, but require distinctly different proofs to support their assertions.

205

15

These conflicts have the potential of abusing the owner's budget, delaying the occupancy and use of his project, and adding substantial administrative costs.

The Difference Between a *Delay/Disruption* Claim and a *Construction Defects* Claim.

Through 40+ years of assisting owners, contractors, architects, sureties, banks and others to resolve disputes that grew out of a building projects, friends have frequently asked:

> *"Oh! You're working on a construction claim? You mean, like a leaky roof, or cracks in the sheetrock? I have a friend whose flooring...."*

A delay/disruption claim is an altogether different animal from the typical construction-defects claim.

The owner most often initiates the defect claim. The owner has a general idea, or maybe even a specific idea of what's wrong. He hires an expert, or several, to identify the cause or causes, and to propose solutions and then brings a claim against the parties who appear to be responsible for the problems and the cost of repairs.

This type of claim can be costly, but the successful owner can often recover the costs of bringing the claim.

But the Delay Claim Sneaks Up When the Owner's Not Looking

While the owner was doing his thing, others on his project were agreeing to disagree.

The causes of a *delay/disruption* claim are usually hidden. Hidden in the actions taken over several days or weeks by multiple participants behind closed doors, out in the field, in manufacturing shops, in field offices.

Few who have not been through the experience understand the implications, the sources, the costs and the frustrations associated with a construction delay/disruption claim.

Even those who have survived a full term construction litigation have little opportunity or inclination to analyze how the claim came about, or how it could have been avoided. Most participants just want to get away from the circumstances and the people surrounding the claim and forget the whole experience.

Even if they tried to understand what happened, the distortions to which the facts have been subjected would make the effort useless anyway.

As a result, even owners who are involved in continuous building programs—cities, counties, districts, universities, hospitals, expanding businesses—invite the same exposures to conflicts and claims over and over on successive projects.

Seeing how a delay/disruption claim comes about, how many months it takes for it to go away, how much money it will consume, and the impacts it will have on an owner's productive time are all insights that should encourage an owner to invest every effort—personal, professional, political—to avoid those conflicts before they take place.

Delay? Disruption? What's the Difference? How Does Each Relate to Time?

Follow along here for a few minutes and be prepared to thumb backward and forward to some other chapters. You will know more about the similarities and the differences these two abuses of time can have on your budget than most arbitrators do. You will be equipped with the types of challenges that should be raised when you're confronted with either or both types of claimed impacts.

They're both the result of some party on the project having abused time. They're usually ganged together in a construction complaint, whether or not they occurred that way in fact. A time-based construction claim is usually a grab bag of every possibility—delay, disruption, acceleration, idle equipment, escalation, loss of business opportunity and the rented field office fax machine (the equivalent of the kitchen sink).

Delay asserts *uncompensated extension of the contract time.*

Proof of the delay-claim assertion requires establishment that there was a valid path of "controlling" or "critical" work that was delayed due to conditions beyond the control of the contractor.

15

Proof of "entitlement" (see the next few pages) for a delay claim requires that the contractor establish that: [1] there was an "approved" or "accepted" schedule in place when the event occurred, [2] the work delayed was "controlling" or on a "critical path" in that schedule, and [3] the cause(s) of the delay were beyond the contractor's control. (For an explanation of "controlling work," see the description of CPM in Appendix A-2.)

Establishment of "damages" for a delay claim requires that: [1] the contractor establish the number of days the work was delayed, [2] there were no "concurrent" (see Appendix A-2) causes of delay for which the contractor was responsible, and [3] the daily rates by which he multiplies the delay days are supportable (which requires various cost reports and formulas to demonstrate field- and home-office overhead rates).

> The concept of *days* in the context of delay claims can be confusing until you get a handle on it. The contractor usually plans his work on the basis of *working* days (5 days/week). Liquidated damages (the daily amount the owner will charge the contractor for every day the project goes beyond its contract duration) are typically calculated in terms of *calendar* days (7 days/week). A fair rule of thumb (it's still necessary to take holidays into consideration) is *"calendar* days equal *working* days times 7 divided by 5."

> The contractor's overhead is usually calculated in terms of *calendar* days, because most overhead costs are accumulated on a monthly basis (jobshacks, field superintendent salary, insurance, home-office expense, etc.). However, some direct field costs can be on a *working*-day basis if they include personnel paid on an hourly basis.

Disruption asserts that the contractor was *unable to use time efficiently*.

Proof of entitlement under a disruption claim does not require the contractor to establish that there was a "critical path" or that impacted work was "controlling."

A disruption claim requires that the contractor prove that he had a reasonable estimate of costs and a reasonable plan for the performance of the work. As you will see in chapter 17, this isn't as straightforward as it sounds. There is often little or no correlation between the costs in the contractor's original estimate, his budget, his cost control and his payment request form. As a consequence, the contractor who's challenged to provide his original budget for work he claimed was impacted can have some difficulty establishing the basis against which the disruption is to be measured. There are usually several hidden factors involved.

The sequence in which the work was actually performed is often different from how it was scheduled. The contractor is often unable to demonstrate similar work on which he was 100% efficient (the "measured mile"), so comparisons to other work are often compromises, or jerry-built out of whole cloth.

Few schedules are maintained accurately enough to reflect actual progress on the project.

The contractor must establish that he (and/or his subs) did not contribute to the reduction in efficiency. Apart from the difficulty of establishing the labor hours and equipment under his original plan, it is often problematical for the contractor to demonstrate that he was not at least partly responsible for the changes in performance and efficiency.

The contractor frequently has used different resources (labor, equipment, etc.) from what he planned, so it can become difficult for him to establish the *amount* by which he was forced to be inefficient.

Apart from whatever differences he claims between his "basis" and the "actual" costs, he must prove that those differences were caused by conditions entirely beyond his control. Contractors and their subs often change sequences of work or change support equipment. They even may have shifted labor forces around on the project.

Presented with these difficulties, together with their own limited perceptions about proof, and confronted by challenges from the owner, contractors often turn to construction-industry guidelines and tables that assign *percentages of loss of efficiency* under a range of conditions. These conditions include *difficulty of performance due to difficulties from stacking* (crowding) *of trades, acceleration, morale, learning curve, ripple,* etc. The ranges of efficiency losses are typically scaled from 1 to 5, equivalent to, say, 5% to 30% losses. The contractor's first challenge is to establish into which range his *impacted* work falls.

15

Contractors will attempt, naturally enough, to apply as high a factor as they believe they can sell. And although the factors presented in these manuals are not necessarily cumulative, it is not untypical for them, or their attorneys (with limited understanding of the conditions that brought the asserted disruption about), to add percentages that are not necessarily additive.

Careful reading of these manuals reveals that they contain many qualifications regarding the rates presented. An effective response will take these qualifications into consideration.

Because of the difficulties inherent in establishing both the entitlement and the damages related to a disruption claim, contractors will often elect to establish a delay claim instead. They will maximize the *delay damages* attributable to the owner (by asserting maximized durations on "delayed" work—see comments on the "collapsed schedule analysis" in chapter 16) in an attempt to recover all their losses under "delay."

In defending a delay or a disruption claim, it is important to distinguish between the causes and their impacts. They should be challenged for what they are, with all their respective requirements for proof.

None of this is to say that disruption claims cannot be valid. Many have been established for large recoveries.

But the owner owes it to himself, his organization and his budget to know the differences. The appropriate response can substantially reduce the owner's exposure to additional costs.

For an owner, the bottom line is that: claims based on delay, disruption or acceleration (accerleration often follows either delay or disruption) are the result of time abuse. Because of their complexity and number of components, an owner will be required to invest substantial time, effort and costs in responding to them—and will still have no assurance that triers of the facts will have the patience, perception or commitment to get at the "truth." It's much easier for arbitrators and mediators to accept the absurd conclusions presented in "analyses" they're not equipped to assimilate.

Piece of Junk...

The contractor had a hired gun with a reputation for smooth talk and a quick graph.

The hired gun had taken the contractor's original CPM schedule—which had its own inherent limitations—and had superimposed on it, from any documents he could lay his hands on, every complaint, presumed delay, contested issue and asserted impact he could invent. And even added as impacts, work that had been discussed but never incorporated into the project.

By the time the delays he had identified were superimposed on the original schedule, the overall duration of the project was about 2 months beyond its actual completion date. He asserted with a straight face to the arbitrators that this was only a "hypothetical" projection that was the result of the multiple delays on the project.

He then gratuitously volunteered the amount of delay for which "his analysis" demonstrated that his client, the contractor, was responsible (about 12% of the overall delay time); and deducted this time from the overall duration. The result, he said, was a "collapsed as-built analysis" of the project. It was presented in a series of dramatically annotated, computer-plotted, color-coded summary charts.

The arbitrators provided these charts the respect they would accord to the Rosetta Stone and gave the consultant their undiluted attention.

When the opposing consultant testified, he attempted to educate the arbitrators to the fact that superimposing actual data on a schedule that was probably only representative of the actual performances on the project for about a week and a half, was inherently absurd and irresponsible. He described the charts as "pieces of junk." At that point, he was interrupted and summarily chastened by the attorney-moderator of the panel. The legal stenographer was instructed to strike his description from the record—and he was advised to restrict any further criticism of the charts.

The arbitrators had found something that was clearly undecipherable by any but an expert, so it would provide an unchallengeable basis for their decision.

Since claims involving time-abuse are almost invariably the products of defective design and deficient contract administration, they're avoidable.

15

It becomes a choice of investing in the front end, instead of the back end, of a project.

As the oil-filter guy used to say on early TV when he reminded us of the difference in cost between an oil filter and an engine overhaul: *You can pay now—or pay later.*

Genesis and Evolution

The contractor submits a Request for Information (RFI).

The architect doesn't respond for 20 days.

When he does, it's with an inadequate answer.

The contractor and the architect argue. The architect denies responsibility for asserted design errors. The contractor asserts both delay impacts and the need for a change order.

The owner's "construction manager" (CM), demonstrating [1] his inability to see problems coming, [2] his lack of experience and objectivity, and [3] his inability to manage problems or people, tells the contractor they'll deal with it later.

The owner, believing what he's been told by his architect and his construction manager—that the contractor is out of line—folds his arms and digs in his heels.

The contractor directs his attorney to file a claim to recover impact damages. The complaint alleges defective design, among other administrative failures ascribed to the owner, his architect and his CM.

The architect's professionalism has been attacked, so he will have notified his E&O carrier. The carrier brings in its own troops: legal department, consultants, accountants. The contractor's surety and its attorney make an appearance. That brings several other attorneys out of the woodwork, representing respective subcontractors and suppliers.

The owner's lawyer—the one who OK'd the original contract terms—finally acknowledges his limitations and advises the owner that he should retain a *construction-knowledgeable* attorney (make sure it's not an oxymoron this time).

The Camp Followers

Before the owner is aware of it, possibly even before construction starts, the contractor may find in your documents the basis on which to build a claim.

He puts his project manager or his cost estimator to work on it.

Then, because the schedule he gave the owner was, likely as not, an ersatz bar chart made to look like the critical path schedule the specs called for, the contractor hires a claims consultant to build a claim. He needs some basis on which to assert "delay," "disruption" and "extended overhead costs."

The contractor calls and tells his attorney—who's been on the sidelines waiting for the next phone call—of his intention to bring a claim and asks him to look over the disputes clause to ensure that he can meet the conditions. Once the claim is filed the attorney or the contractor will retain an "expert" to analyze the issues.

The expert looks over the schedule, meets with the contractor. and helps expand the issues. He will start to assemble an "as-built" record of the project-claims issues.

If the participants can't get down to talking calmly about resolving issues, other camp followers will show up. The architect will retain experts to absolve him of his errors. Some of his subconsultants will do the same. Since several disciplines are involved, the owner will start to see consultants appear on the scene that he never had a chance to meet during the design process.

The contractor brings in one or more claims consultants. Subcontractors, either cooperating with the contractor, or directing their complaints at him, assemble their own squads of estimators, consultants and attorneys.

Delay issues will require the services of "schedulers" for all sides. Damage issues will require the care and feeding of more experts to analyze costs. They will arrive with truckloads of documents and opinions.

The owner's new partner—his "construction-knowledgeable" attorney—will advise him that he needs to retain equivalent experts in order to insulate himself from liabilities, or at least to recover costs from one or more of the other parties he retained on the project.

15

And each stage of the claim's development and defense will bring more participants to the party: interrogatories—*paralegals*; depositions—*associate attorneys*; discovery—*more associate attorneys and legal assistants.* And document production—*which will bring document clerks and copying costs*—hasn't even started.

Nor have the costs of the court, the arbitrators, the mediator or judges-for-hire been considered. All of these *dispute industry participants*—attorneys, panelists, consultants, mediators—will bring overhead costs and staff assistants with them.

All of them—the "essential strangers" and "other strangers" as well—will be focused on exploiting, each to their own advantage, their claims on the owner's project and resources. The "strangers" that you believed you were getting to know will soon start looking more like leeches.

Leeches are difficult to remove. Each has an idea how much blood it must draw before being removed from the delicious corpus of the owner's project. FILO usually applies (first in—last out).

The Components of a Construction Dispute

"I promise I'll be there at 3:00."

You know at 3:40 that somebody's failed to keep his or her promise.

It would be convenient if broken promises on a construction project were as easy to track. They're not. But there are rules.

In order for one party to a construction project to establish that compensation should exchange hands for a broken promise, that party has to prove two separate components of its claim.

The claiming party has to prove it's entitled to something, and to prove the discrete dollar amounts of damage it has experienced.

1. **Entitlement:** *The right to benefits… which may not be abridged without due process.* The proof of the pudding. The right to recover once one party establishes the merits of his assertions. The right a party to a construction contract acquires if he, she or it is able to prove that the terms of the contract were breached by the other party. As an element of this proof, the claimant also has to establish that he has met the requirements for his own performance to the extent that they represent conditions precedent to the other's breach.

214

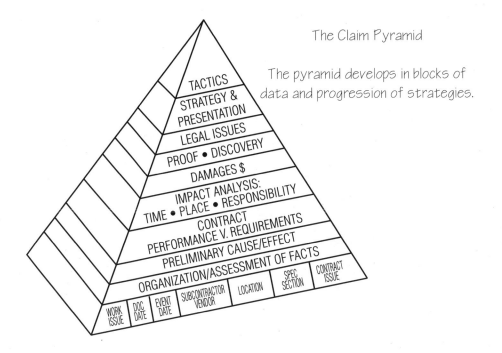

The Claim Pyramid

The pyramid develops in blocks of data and progression of strategies.

For instance, for me to prove you didn't perform, I have to establish that I didn't interfere with your ability to perform.

2. **Damages:** *The amount of compensation a party claims it should recover once entitlement to that recovery has been established.* The effects of the causes about which entitlement has been established. Usually, the law wants to see specific damages associated with specific causes, although this rule has been relaxed considerably as the complexity of relating effects to causes has increased.

Even though entitlement has been established for the recovery of damages, however, an audit of the claiming party's accounting or cost records will usually be required to verify the quantum of claimed losses.

By the time issues of entitlement and damages, causes and effects have been joined, several new participants have been introduced into the disputes process.

15

The Time It Takes For A Claim To Get To Arbitration

Action	Cal. Days	Cumul. Months
Claim is filed	1	0
Response is filed	30	1
Communications between attorneys	60	3
1st set of interrogatories	45	4.5
Response to 1st set of interrogatories	30	5.5
Respondent set of interrogatories	15	6
Response to respondent interrogatories	60	8
Production of documents	120	12
Continuing discovery—depositions	120	16
Trial setting conference	30	17
Mandatory settlement conference	45	18.5
Mandatory judicial arbitration	45	20
Set arbitrators and arbitration dates	60	22

Along with the need for an attorney to interpret the legal aspects of the contracts and perfect the claim under the terms of those contracts, one or more experts will be required to assess who did, or did not, do something when; to whom; and what conditions in the construction documents or in the industry's recognized standards of performance were elements of the asserted breach of contract.

One or more of the parties have brought in cost consultants, estimators or accountants, or all of them, to assess the amount of damages.

All of them need a considerable collection of documents to make their assessments. The costs and the time for resolving the dispute have only just begun to accumulate.

216

The Time and Costs It Takes
for a Delay Claim to Go Away

Your project representatives recommend that you reject the contractor's assertions for extra costs and extensions to his contract duration.

With the issues joined, the contractor's attorney starts a legal battle that consumes another 22 months before the matter even goes to arbitration.

Then it's likely to take another five months, with recesses for the arbitrators to fit a hearing of the dispute into their schedules, vacations and other matters. If the interruptions in a court trial are frustrating, consider spreading the story of a $3 million dispute on a $12 million project over five months. By then, "facts" of the matter will have become highly fragmented and disjointed. And once the case is submitted, it will take eight more weeks before the arbitrators render their decision.

Whether the arbitrators' decision splits the baby in half, makes a substantial award to the contractor, or finds in the owner's favor, you will likely never know what moved the arbitrators to their decision. They have no obligation, usually, to disclose either their reasoning or their decision[1].

When it's all over, you may try to add up the costs, or you may just go back to your usual activities. Either way, few owners will ever know the total costs of delivering deficient construction documents to contractors.

The Ultimate Irony: Wasting Policies

Everything has been shaken out. The owner has been directed to compensate the contractor for the impacts of design defects. But the owner's architect has in the same action been found responsible for those design defects. It's then that the owner discovers the final irony.

When he negotiated his design agreement, the owner made sure there was an errors and omissions (E&O) policy to protect against this sort of contingency. But, somehow, he missed the fact that it

---

[1] *Though the AAA, in an effort to recover its lost credibility, has begun to recommend that arbitrators explain their decisions, the process is spotty, inadequate and far from consistent.*

15

was a "wasting" policy. The face amount of the policy has been reduced by the funds paid out by the architect's E&O carrier to defend the claims made against its client's performance.

Most of the money the owner expected to recover from the architect is gone—into the pockets of the lawyers, the consultants, the mediators and arbitrators—paid out by the surety to protect the party who was responsible. The same one who provided that E&O policy to the owner.

> *The $500,000 E&O policy from which you thought you'd be indemnified for claims against defects in your construction documents is down to $72,642.17.*

The owner is limited to collecting what remains of the coverage in the policy, plus whatever he can recover directly from the architect and his consultants—a recovery effort that will require the expenditure of additional legal costs.

The Era of Disputes

As construction management grew, so did the incidence of construction claims. From 1965 to 1995 the American Arbitration Association (the AAA) experienced a 482% increase in the filing of claims for arbitration.

Over the same 30 years that the AAA experienced its own increase in caseload, a growing number of firms in the construction industry were discovering opportunities to exploit the resources of owners.

They were encouraged by attorneys who widely promoted seminars and workshops to explain to subcontractors, suppliers and others their rights to bring their own claims. The parade was quickly joined by a burgeoning cottage industry of scheduling and construction claims gurus.

The increase in filed claims led to increased caseloads in the courts, more arbitration and an exponentially expanding number of alternative disputes resolution (ADR) arenas.

While this was happening in private construction, government disputes review boards experienced their own exponential increases in caseloads.

It's been estimated that the number of construction disputes that got as far as the formal filing of a claim increased at least tenfold between 1965 and 1995—1,000%.

This increase is only what's formally recorded. For every filed claim, several times that number of projects overran their budgets by more than 10 percent. Every year, thousands of projects experience cost overruns, change orders and disputes caused by defective design documents, ineffective administration and poor budget estimates. Most of them are compromised without being formally recorded.

Even though the costs of settling these overruns and compromising these disputes is usually less than the cost of going full term to trial or arbitration, it is a substantial drain on an owner's time, resources, patience and productivity. To say nothing of the delays and administrative expenses associated with resolving most disputes.

The unfortunate fact is that the completion of a building project within its original budget and planned duration has become the exception rather than the standard in the construction industry.

7

·HOLLOW PROMISES AND MORE ABSURDITIES

In the construction industry, success is not the product of buzzwords, but of careful, relentless planning and follow-through.

It's a matter of anticipating the interactivities of the strangers—the architects, contractors, construction managers, sureties, inspectors and lawyers—who will come onto your project to crowd and jostle you.

The standard contracts and panaceas the owner will be offered do not provide the tools with which to anticipate, measure or document these interactivities.

16

PANACEAS THAT AREN'T

Panacea: A supposed remedy, cure or medicine for all diseases. The Greek goddess *Panacea* was supposed to have the answers for all problems. It's appropriate that she was only a figment of Greek mythology.

But construction is a peculiar environment. It's easily exploited by oracles bringing panaceas to cure its ills. Oracles who will help you *partner;* who will help you *avoid claims;* who will help you achieve *Total Quality Management.* The buzzwords they promote already have the ring of success because they've worked in business and industry.

But if you look inside, many of them are Trojan horses.

The problem common to the application of these panaceas in the construction industry is that they're imported *without* the conditions that made them successful in business—conditions that are only infrequently possible in the construction environment.

One shortfall of these panaceas is their typical failure to identify a clear leader, one who understands that it is the duty of leadership to identify clear goals, responsibilities, scopes of work and standards for measurement.

If these buzzword ideas are going to work in construction there are conditions-precedent that the owner—the leader—of a construction project needs to put into place.

They include [1] a clear statement of goals to be met (Program), [2] an effective system of administrative procedures and contracts (Plan), [3] competent tools with which to measure the performances under these contracts, [4] clear, buildable construction documents, and [5] consistent and responsive leadership.

Construction Management (CM)

CM started as a good idea. Someone independent of the design team was retained as the owner's representative to manage the whole process of design and construction. This idea was a di-

223

rect evolution of integrating design and construction into *fast-track* contracts for military development during the Cold War. (*Fast-track* and other types of contracts are defined in chapter 19 and in the glossary.)

CM No. 1

You may have wondered why, in earlier chapters I've frequently put certain phrases, notably "construction management" in quotation marks.

It's because "construction management" has been so ineptly defined and performed since it was originally conceived that it frequently delivers much less than it promises.

The Cold War was in progress in the 1950s. The U.S. Government was in a swivel to develop a defense establishment. Contracts went out *fast-track, design-build* and T&M (see chapter 19). The government recognized its designers' and its own inability to manage such contracts. The government invented the concept of "construction management" to bring a firm with management experience into the process early to coordinate integrated procurement, design and construction.

Even though the government wasn't always effective in making up its collective mind about what it wanted, or in controlling its "budgets," the idea did accomplish the accelerated delivery of its Cold War inventory.

There was substantial value in the original application of the concept. It recognized that *management* is a unique discipline that can deliver significant benefits.

But since then all manner of firms in the construction industry have exploited the buzzword without defining or consistently delivering the values inherent in effective management. (The end of chapter 7 recounts some of these values.)

CM No. 2

In 1968, the Associated General Contractors of America (the AGC) hoisted its sails and pushed off into uncharted waters with a seminal pamphlet.

The AGC is a directed, well-organized association, reflective of the major firms in the construction business. As a group, contractors are no-nonsense types. The AGC broached no competition with its initial sally into the seas of construction management.

In 1968, the AGC's no-nonsense definition of what the government had invented was:

"The construction manager is that general contractor who..." (The first line of the AGC's pamphlet.)

The AGC then went on to reformulate a body of services that the CM would provide for an additional fee. In retrospect, it's difficult to distinguish some of these services from many that the owner believed he was supposed to receive through other contracts.

Comprehensive Services

The AGC had been on the high seas for a year. The American Institute of Architects (AIA) convened in Philadelphia in 1969 to clabber together a ship to catch up. *We, too, can be managers of construction.*

The AIA's answer was "Comprehensive Services." *We can do anything you can do. All we need is to print some new business cards like you AGC guys did.*

The AIA's ship sank in sight of port.

The Team Concept

With a failed "Comprehensive Services" on its hands, the AIA started to promote *The Team Concept. We're all in this together, so let's not become confrontational. Give us the assignment to put the Team together, and we'll show you how a project should be managed.*

This attempt to fill the vacuum that resulted from the contractors' assumption of the management role was around for a very brief time. The industry (but not necessarily the owners) quickly recognized that this buzzword was also stillborn. It has disappeared over the horizon with "Comprehensive Services."

Its demise resulted essentially from the fact that *The Team Concept* was not clear about who the Leader was, since no new abilities had been added to architects' management competence.

16

Today, members of the AIA promote their own construction-management services. (So does everyone else—see CM No.3, following.)

Partnering

The central theme of *partnering* is cooperation. An excellent idea.

The concept is borrowed from industry. There it makes sense. The members of an R&D or production team have a common goal. They "partner" their talents to produce that goal. The partnering process is enhanced by the fact they work in the same environment—a laboratory, a set of offices—where they share equipment, communications systems, common working environments, and readily maintained schedules.

What makes the direct transfer of the concept to construction an improbable dream is that the strangers the owner hires usually don't have the motivations or the continuing relationships to become amalgamated and balanced as do the "partners" in business and industry.

The other inherent contradiction in *construction partnering* is that it's usually invoked after the owner's goals, the design and the contracts are in place. It's already too late to bring all these diverse participants together if a clean set of construction documents hasn't already facilitated it.

Partnering is typically implemented by bringing in a "facilitator" after the construction contract has been awarded, to explain how everyone will benefit by working cooperatively together. A good idea. (Not that it should take a "facilitator" to demonstrate it.)

> The "facilitator" demonstrates competence in construction and in people skills. He/she puts the parties into challenging "what if" situations and demonstrates what some attendees may not have realized before—that problems can be worked out without loss of life and limb.

> Mechanisms are put into place that will raise problem solving to the next level of management on a timely basis if lower levels are unable to find solutions.

If the lessons stick and the mechanisms are honored, the partnering sessions can benefit even those projects burdened with defective design, within the limits of the time-honored practice of

226

horsetrading that knowledgeable and mature people employed effectively in construction up until about 30 years ago—a time when architects and contractors communicated more readily than they do in a litigation-littered industry today.

The idea of "partnering" can be productive if the participants are committed to openness, candor and a willingness to work out problems. The basic idea—cooperation—should be encouraged. It's unfortunate it's saddled with an absurd name that suggests equal sharing of motivations and exposures. Partnering—or horsetrading—can benefit a construction contract only as long as the owner and his architect are prepared to recognize real deficiencies in the documents and to respond reasonably.

But the owner who believes that a *partnering charter* will accommodate defective documents without allowance for reasonable compensation to contractors is in for one of those surprise parties we discussed earlier. It's absurd for an owner to believe that all these "partners" will continue to cooperate with the owner or his architect if the owner and architect go into denial and if those documents start to compromise those contractors' own sharply felt motivations.

Lunch with a few subcontractors will tell you how tenuous are the commitments to the *partnering charter* that attendees sign when they complete partnering sessions. The tenuousness of these commitments is not so much a reflection of cynicism as of accumulated experience—and the *above the fray* responses of some owners and design consultants.

If things do start to get churned up in the fan, the first thing that's shredded is the partnering charter.

Partnering works particularly well on engineering projects. On such projects (highways, bridges, heavy-construction projects) the team is usually organized around the management of the contractor who will integrate the design and the construction of the project. There, *partnering* is a natural. It's facilitated by the fact there are fewer contractors on an engineering project than on a building project (divide by 5 or 6), and by the fact that many of the contractors and designers on engineering projects have

16

worked together before. All of which is complemented by the fact that, on average, engineers are better managers than are architects.

It's worthwhile noting that the kinds of projects on which *partnering* works best tend to be "horizontal" ones on which both time and resources can be adjusted. These types of projects can often be divided into elements that can proceed concurrently (different sections of a freeway, for instance) and built with separately integrated crews.

Total Quality Management (TQM)

Total Quality Management is another concept borrowed from business and industry, albeit U.S. industry first had to send it to Japan before recognizing its value.

Like the other panaceas here, TQM requires clear management direction, continuity of association, balanced motivations, and effective management tools for measurement and direction of goals.

Partnering: Unbalanced Motivations; Unbalanced Liabilities

A partnership is supposed to be based upon balanced liabilities of the partners. The owner is on the line for the funding, the maintenance and the operation of the project over the next 30 to 40 years. Those he hires expect to make a profit on his project, and move on.

Hardly a balanced relationship.

If the concept means anything to an owner, it should invite the owner to become his own facilitator and to facilitate a "partnership" during design, when it will deliver the optimum value.

This is the core idea of the B/BA that will be central to your Project Plan. The open, analytical and candid relationship the B/BA invites should facilitate a balanced working relationship between the owner and the design consultants that will minimize the need for horsetrading later.

If the owner can't establish a partnering charter with his design consultants, then he may have the wrong architect.

All of which makes TQM in construction just a platitude, unless the owner has developed a well-conceived Project Plan for project development.

Plan Check

A Plan Check is typically a review of the construction documents to ensure that they comply with applicable codes and building regulations.

If that's what's delivered with a Plan Check, it has value.

The important thing for an owner to understand is that a Plan Check does not confirm [1] that the documents have adhered to the original Program, [2] that they can be built within the budget, [3] that they are even buildable without modifications, nor does it [4] warrant that they're internally coordinated and without errors, ambiguities and conflicts.

Plan Checks are typically undertaken late in the design process, somewhere around the 75% - 90% Construction Documents phase. As such, they come along too late to add any but a reactionary overview of the documents.

Value Engineering (VE)

The purpose of Value Engineering is to incorporate into the project less costly or more efficient alternative materials or equipment without demeaning the value of the project.

The application of VE has been extended to the construction phase in some contracts. Contractors are invited to make recommendations early in the construction process that will benefit the project.

Value Engineering is an excellent concept, properly invested into the project. In fact, it's a mindset the owner should motivate from the outset, not only for first costs of the project, but to optimize the maintenance and operating costs of the project over time.

It should become part of the owner's programming criteria and an element of the progressive B/BA implemented throughout the design phases.

Hospital Project: The CM's "Value Engineering"

The hospital's new nursing tower was budgeted for $7 million. The architect and contractor-construction manager (CM) team would work together to produce the documents. The CM would be paid $20,000 for its "value engineering" efforts during design. Both were hired on standard AIA contract forms.

With the Design Development documents completed, the CM provided a Guaranteed Maximum Price (GMP) of $7 million for construction.

Two and a half months later, the construction documents were complete. The contractor-CM was ready to start work—and told the hospital that the cost of the project would now be $9 million.

The hospital had no option but to proceed with construction. However, the board was very concerned about what had happened and what might happen with the balance of the project. It brought in an independent construction consultant to review the work-in-progress and to review the accumulating documentation and the CM's accumulating change orders.

The hospital learned that:

- The interstitial space (the additional space provided between floors to hold mechanical and electrical equipment) was so limited that it would not accommodate all of the equipment.

- Neither would the interstitial space hold the ductwork that was to run in it. Ceilings would have to lowered in the occupancy areas to allow ductwork installation there.

- As work progressed, it became apparent that medical procedures conducted in rooms fronting on the street would be reflected to the street because of the improperly located sloped windows.

- No backflow-preventers had been incorporated into the facility's plumbing system: a code violation.

- The changes for which the CM was requesting additional compensation were for basic design errors in the documents.

The architect and contractor-CM had obviously opportunized on their own long-term relationship from past (and, potentially, future) projects to attempt to exploit the owner's naïvete. The consultant recommended that the owner bring an action to recover its excess costs, along with the $20,000 VE fee for which it had received no value.

In the lawsuit that followed, not only were the architect/CM held to their original $7 million GMP, but they were additionally required to correct, at no further cost, the deficiencies in the project.

CM No. 3

Today, you can pick up the Yellow Pages of the phone book in any major city and under the "Construction Management" find cost estimators, CPM consultants, inspectors, plumbers, landscapers, architects, general contractors, claims consultants, accountants, home builders, equipment suppliers and kitchen remodelers.

But if you read closely the contracts these firms offer—including the standard AIA and AGC forms that incorporate CM—you will find that they are [1] entirely deficient in their definitions of work scopes, [2] porous in their delineations of responsibilities, and/or [3] redundant in terms of the fees and services in the other contracts on the project.

Construction Management firms are good, bad, ugly and worse. The owner's challenge is to eliminate the charlatans and the incompetent down to the core of firms that will assign a knowledgeable and mature manager who will represent the owner's interests with objectivity and integrity.

Program Management

As implemented by two different firms on different projects, Program Management included:

1. Comprehensive management of the entire development process, from the initial planning, to direction of activities for occupancy of the completed project.

2. Development of a CPM schedule, plus monthly updating as reported by the contractor solely to meet the requirements for payment.

Two definitions for Program Management, about as far apart as you can get.

Program Management is another buzzword that has lost its meaning through exploitation and loosey-goosey acceptance of whatever's offered. The owner must take the initiative to define the scope of work for the services he or she requires.

16

Claims Avoidance

This empty phrase is promoted as though it were a special effort.

The avoidance of claims cannot be guaranteed. Reducing the potential for claims, or the liabilities associated with them, requires effective leadership and management.

Constructability Analysis (CA)

As with all the concepts above, this one will add value to a project in relation to how well the scope of work it contemplates is defined, and how well it's implemented.

To be of real value, a Constructability Analysis (CA) must be comprehensive; must be implemented through experienced, objectively motivated analysis; and needs to be undertaken progressively as design advances.

To be comprehensive, the CA must cover the content of the Contract Documents in their entirety—the agreement, the bid documents, the administrative specifications, the technical specifications *and* the drawings, plus the owner's Plan for exercising leverage and taking action when measurements demonstrate the need for action. That includes the delineation of the procedures for reporting, measuring and documenting performances of all the parties under contract.

The progressive review and evaluation of design needs to incorporate a VE effort, a Plan Check and a progressive refinement of construction costs and schedules until, by the time the construction documents are turned over to contractors, the project has already been *built on paper.*

It's because this standard is seldom met that I've coined the term Buildability/Budget Analysis as an interim name to distinguish it from the typical CA.

Value-Added

Value-Added is the contribution to value that results when the separate components of a product are brought together and the end product is worth more than the sum of the pieces that arrived at the other end of the process.

After the Barn Door...

A Southwest city recently announced it was terminating its CM, paying the firm $2.6 million to go away.

The firm had negotiated a contract with the city by which it was to be paid 50% of the first $10 million in savings based on the bids received on a $120 million convention center, and 15% on additional savings. The bonus, however, *was not tied in any way to completion costs,* and the CM had no liability for cost overruns.

Having entered into this agreement, the city then found itself in a dispute with the CM about scope and the staffing of the project.

The city acknowledged that it had learned a lesson: *(on future projects) "We must make sure we have the experience, if not on staff then on a consulting basis to make sure our interests are represented."*

What the owner of a construction project needs to do is to consider whether the addition of another participant to his project will contribute to control of costs, to the administration of the schedule and to overall efficiencies in everyone else's performances.

To make this judgment, the owner first has to know what the respective duties of the parties are, and which duties and analyses will be more objectively administered by a party independent of both the architect and contractor.

Inherent in making this determination is a clear picture of what the owner's own duties will be during design and construction, and how he can be effectively and comprehensively advised so that he can perform those duties in a timely and decisive way.

Then, he has to determine whether the applicants he interviews really are independent of the other parties; have the competence, experience, objectivity, maturity and clarity of mind to articulate those duties; and are committed to objective representation of the owner's goals.

16

The owner also needs to have an essential "chemistry" with the representative he hires in order to be able to rely on the consistency of his representation over the life of the project.

Once he decides this, he can start to define the representative's role: how much, and what kind of services during the design phase? What services during construction?

We've already discussed that it takes a certain combination of experience and art to manage the multiple resources required to construct a building project. (See chapter 7.)

These skills need to be applied on a continuing basis. A lot of paper flows between the contractor and the architect; the contractor and his subs; the architect and his consultants; and among all of them and other parties—more paperwork than most owners have the interest, the time or the experience to evaluate themselves.

So, it's a foregone conclusion that the processing and organization of those records is something the owner needs to delegate.

The value added an owner is looking for is the insight required to undertake this evaluation and to recommend effective actions when they're required.

The owner cannot afford to delegate the evaluation of those records to someone who does not have the competence or the motivation to employ those records objectively on the owner's behalf.

A Real Panacea

There is no substitute for being able to work with a pencil and calculator through the formulas behind the computerized reports produced on a construction project.

Do you remember the satisfaction you had the first time you wielded a new Ticonderoga pencil?

Once you add two more satisfying materials to a Dundee marmalade jar of pencils—a large Pink-Pearl eraser and a role of scrim paper—you're likely to find a simple and productive satisfaction in picking them up again.

That's not to discount the usefulness of an adequately loaded personal computer. But you'll find that the responses to most of the absurdities above will arrive first out of the end of a sharpened pencil.

Value Added?

An $18 million bond issue was approved on one school-modernization pro-
gram that involved several campuses.

Of this amount, it was demonstrated by independent analysis that not
less than $11 million of the funding would have to go into utilities, site
improvements and basic distribution systems. This analysis demonstrated
that, carefully packaged, the civil-engineering work could be accomplished for a fee of approximately 6% for both the engineering design
and construction: $660,000.

Over the years, however, the state department of education had been inten-
sively lobbied to establish guidelines that mandated [1] all school design
work had to carry the stamp of an architect, and that [2] the guideline architect's
fee for renovation work would be 12%.

Although the architect would do little but put his stamp on the work of the
engineers; and although there were options available under which the district could award the utilities packages separately, the architect prevailed
with a timorous and naïve administration (with the school board essentially
uninformed) and was awarded the full 12% fee on this $11 million of work.

The architect's fee on this work totaled $1,320,000. It could have been
designed *and built* for about $660,000. The district wasted more than
$660,000 that could have gone into books, computers and other im-
provements in its classrooms.

Value Added?

17
EMPTY DATA, OBFUSCATORY BUDGETS, ROARING MICE AND MORE ABSURDITIES

Peter Drucker recently reminded us (again):

"It is all too easy to confuse data with knowledge and information technology with information."

Mr. Drucker wasn't speaking specifically about construction, but if he wanted to find a paradigm proof for his statement he could have stopped here.

As if the panaceas in chapter 16 didn't offer enough hollow promises, plenty of other absurdities rear their heads in the course of project development. You won't be surprised to hear it repeated that project-status reports on any project can be manipulated and can be full of empty information.

But construction reporting has been raised to a manipulative science encouraged and facilitated by sophisticated computer software.

SCHEDULES

We've already established how critical time is in a construction environment.

In spite of that, the construction industry has bought into the notion of the *"CPM scheduler"* as a recognizable, stand-alone discipline.

Is the "CPM scheduler" someone who has accumulated considerable experience with construction and has then added competence with the critical-path method for analysis and planning?

Nope. He or she is someone who has probably taken a three-day scheduling course with the purveyors of a scheduling program and knows how to work a computer.

Whatever experience he has with the logistics of construction, planning, budgeting, or directing construction resources is often incidental, if not non-existent.

17

Would you have anybody but your surgeon schedule the logistics of your heart-bypass operation? Why not the nurse—she's watched several operations? How about the assistant—he's been in the OR preparing the instruments for several operations?

Why not have the cabin boy plot the QE II's course and ports of call? Or the aircraft steward schedule maintenance on the 747 you'll be flying to New York?

How about having the equipment manager direct the defense in the Super Bowl?

But the industry—contractors, construction managers, architects and owners alike—put the instrument that will be used to plan, schedule and measure the single most important resource on a construction project into the care and custody of a 25-year-old junior engineer. All that's expected of his ability is that he knows how to run the computer software to produce the reports called for in the specifications, and to process the monthly payment report. His qualifications are little different whether the project is a wastewater treatment plant, an airport, a hospital or a highway project.

The general contractor, required by the specifications to submit a CPM schedule, develops the logistics of *his own work,* prepares an outline bar chart, and calls in a "CPM scheduling consultant." The consultant (who submitted the lowest price to meet the specs) is hired to convert the bar chart into something that will satisfy the specifications.

The "Critical Path Consultant" needs only the qualifications equivalent to a stenographer to satisfy. He takes the bar chart, makes some notes on what the superintendent says about his work, or about the delivery of structural steel, and adds some lines to the bar chart to make it look like a critical path analysis.

The end product will have 30 to 40 activities, of 10-, 20- and 30-day durations, for the general contractor's work; plus another 25 to 30 activities of 50- to 60-day durations for the subcontractors. Altogether there will be 70 to 80 activities, connected with squiggly lines to represent a "CPM analysis"—for a $4.5 million project.

Does the contractor usually get away with this kind of shallow response to the specs? Yes. Because neither the owner nor the architect understand CPM. And because the "CPM expert" on the

CM's staff was just two rows back in the same CPM seminar attended by the contractor's "CPM consultant."

The superintendent won't bother with the schedule, and probably won't be able to read it anyway. He'll run the job out of his head and with his three-week look-ahead bar charts. What he and his boss want is to be assured that they can use the CPM to get paid each month.

How has such a powerful tool sunk so low?

It has somehow become gospel in the industry that a *critical-path* schedule is difficult to read.

Superintendents, who can read a set of complex construction drawings and interpret the specifications that go with them, find it difficult to to draw lines that represent their plans for the project. There's a perception that a construction-knowledgeable person shouldn't waste his time with it because his time is too valuable.

Then there's a perception that it will only be accurate for a couple of weeks anyway (often true), and it will have to be revised, or "updated." But no one can find the time or energy to rethink relationships that keep changing. It's easier to react *after* things happen.

The shallowness of the thinking (or lack of thought) that went into the original schedule can sometimes become an embarrassment. Add a stubborn and obtuse CM and the whole issue of keeping a spotlight on *time* can become a badly bollixed affair.

The owner and the CM fussed, fumed, rejected and then either finally accepted the original schedule, or tacitly accepted it by allowing the contractor to use it for reporting and payments. If the contractor tries to change it, they fear that something's being put over on them, because they didn't understand it in the first place, and didn't sit down to review it in detail with the contractor at the start of the job. Everything was done at arm's length, in an atmosphere of distrust.

So time gets thoroughly out of hand. No ones knows what the actual status of the project is, or whether it has a true prospect of being completed within the contract-completion date.

The owner has delegated (abrogated?) the measurement of time to strangers with limited motivation to measure or control what can be the owner's most important resource.

17

Then, the contractor wakens to the fact that he has lost time for which he's accountable. He looks around for some way to pass it off, to demand a contract extension from the owner. The CM and owner reject his request. The contractor presents a claim.

The original indifference to the scheduling of the project now takes on new significance. All the absurdities and ironies we've already discussed are set into motion.

Meanwhile, the Mouse Has Roared

When the critical-path method—nee PERT—was invented in the late 1950s, it was immediately recognized as a powerful tool for analysis of the processes and resources required to optimize the relationships of sequential activities.

On the first major project to which it was applied, the anticipated overall duration of the work was reduced by 50%, with considerable savings in resources.

On its second application (its first commercial one), the duration of a repetitive operation that had been performed many times before was reduced by 57% without increasing its original consumption of resources

The tools common to both of these scheduling analyses, and to the schedules that emulated them over the next few years in industry and construction, were lots of sharp pencils, many destroyed erasers and the occasional use of a lumbering computer.

The computer originally facilitated quick and accurate addition and subtraction. Addition and subtraction are the only functions required to calculate the durations and relative dates that result from a critical-path-network analysis. The pulse of the analysis was established by the rhythm of the pencil and the eraser. Idea led to idea. Trial and error led to improvements, and to the shrinking of the eraser.

As the number of activities grew (the first application is reputed to have had 20,000 activities in its project analysis), the computer added and subtracted, added and subtracted, reducing the time of manual calculations, carrying the addition of resources right along with it.

But then came the computer monitor. And color graphics. And the mouse. The sophistication of the computer displaced the pencil, the Pink Pearl eraser—and thought.

It is now possible to produce entire "critical path diagrams" (although the term's a misnomer applied to what's often produced today) entirely on a computer screen. It only requires a mouse and a scheduling software program to produce highly sophisticated-looking schedule listings and charts and "as-built analyses" with no realistic comprehension either of *what will* or *what did* happen on a project.

In the past 20 years, numberless schedules have been produced with endless pages of scheduling data. Buildings were painted before the roof was installed. Some even before the walls were erected. Walls were erected before footings were constructed. Mechanical systems were tested before power was available. Windows were installed before walls were framed.

"As built collapsed schedule" analyses presented in arbitrations have been so absurd as to assert damages for delays to work that *had been deleted from the contract before the work started.*

"Collapsed CPM analyses" have been used to sell arbitrators "impacts" that are blatantly contradictory to any competent reading of project documentation. It's a sad reflection on the ignorance and diffidence of arbitrators, special masters and other triers of construction disputes that they have no idea what these presentations say—and make no effort to obtain the assistance of someone who does.

The mouse has roared—and in the process has scuttled the prospects of what started out as an exceptionally valuable analytical and management tool.

THE LAW

In all its manifestations, the LAW—judges, arbitrators, mediators, special masters and parties' counsel alike—is often beyond its collective competence when it comes to understanding the interactivities required to get a construction project designed and built.

17

In itself, this ignorance about construction is not bad. It requires a lot of concerted effort in the legal profession to keep *au courant* with the codes, regulations, changes in civil law, legal precedents, legal findings and the politicking required.

It would be enough if the profession were consistently effective and objective in its applications of the law.

The problem is that the *lobotomy manifested as ego* that few attorneys have escaped prevents them from acknowledging ignorance of the proceedings in any other profession, including those essential to the construction process. When construction disputes are presented for consideration, pontification often substitutes for objective judgment. There is an amazing disdain for the judgment of others reflected by the legal fraternity, as though the law provides the only true insight to the universe.

There is an amazing disdain for the judgment of others reflected by the legal fraternity, as though the law provides the only true insight to the universe.

Instead of associating objectively qualified judgment to test the facts asserted by the parties, the disputes-resolution process flies right by examination of the facts directly to *resolution by whatever sources of funds are available.*

It is little wonder that sureties are paranoid—not that their own legal counsel contribute much objectivity.

INFANTICIDE—AND FRANKENSTEIN

With regard to the technique known as the Critical Path Method, we've arrived at the ultimate absurdity.

The construction industry has thrown the baby out with the bath water. It has given up on the use of an extraordinarily powerful analytical tool because it's too much trouble to develop and maintain. *It uses up too much energy, too much brainpower and too much time (there's an irony).*

The industry has bought off on simplistic CPM computer graphics prepared by computer-scheduling jockeys who wouldn't survive on a jobsite. It accepts mouse-drawn hollow information made to look like something it isn't, because the color-coded computer output makes it look more sophisticated than it is.

Disputes-resolution arenas, presented with ersatz claims, accept these "analyses" without challenge because of their unwillingness to admit ignorance of the many ways computers can be manipulated to mislead.

So we have one more irony, and one more absurdity. The infant that showed such promise was strangled before it became a teenager, by the industry that adopted it and by the computer that was enlisted to facilitate it.

Now, like a misshapen phoenix, it has reappeared, metamorphosed into a monster that begets some very absurd conclusions and construction-claims resolutions.

CONCEPTUALIZATION

It's natural for an owner to believe that in the process of designing it, his architect has visualized the construction logistics that will be required to build his building.

But architects and their consultants repeatedly produce design details that aren't buildable. Different planes of the building don't come together. Three physical objects are drawn into the same space by different disciplines. Bolts can't be bolted because another assembly that had to precede it is in the way. Dimensions don't work. The modifications required to remedy these discrepancies can be costly and time-consuming.

It's also natural to believe that the construction superintendent visualizes in three dimensions what he finds in the architect's documents.

But a lot of superintendents, despite the fact they work with physical materials all the time, can't, or won't bother to translate those documents into three dimensions. They plan their own work—utilities, foundations, walls—and leave the rest to their subcontractors. If their subs turn up a condition that's unbuildable (or worse, an unbuildable condition is discovered too late to avoid delays), there can be substantial consequences for the owner's budget.

OBFUSCATORY ESTIMATES, BUDGETS AND COST REPORTS

When a contractor undertakes a bid estimate, his first efforts include separating the scopes of work as defined by the specifications. Then, he prepares a *quantity take-off* of the work his own forces will perform. Subcontractors will meanwhile be doing the same with their respective scopes of work.

17

The general contractor then multiplies his historical unit costs for similar work—which may or not be stored in a computer database—by the quantities in his *take-off* to extend the cost of his own work. His next step is to combine the costs he's *extended* with bids received from subcontractors. The process is called *assembling the bid.*

The organization of the bid usually follows the organization of the sections of the technical specifications, because it will be necessary for the general contractor to account for possible gaps and overlaps in his own estimate and parcel out scopes of work among subcontractors.

> For example, when temporary equipment—such as scaffolding—is required to perform the Work, it is essential to establish which contractor will supply it.

The general contractor may "plug in" costs for work for which he receives no sub-bids, or because he believes he can get a lower cost later on. There will be a lot of backing and filling as these costs come together in the last few days before the bid date. The final hours, during which adjustments are made in a frantic atmosphere of telephone bids and judgment calls are too hectic to describe here and have already been described in chapter 4.

It's fair to say that when the sealed envelope is turned in a few minutes before the bid-closure deadline, probably no single person on the contractor's staff has had much of an opportunity to think through the logistics of the project if the bid is successful.

If the contractor is successful, his next step will be to re-assemble all these costs into a *construction budget.*

The general contractor can be both creative and obfuscatory with the construction budget. He may not want his actual estimated costs to go out to the field, to his superintendent, to competitors or to the owner. By rearranging his allocation of dollars, he can set high goals for his own people to meet and allocations that bear no relationship to the actual sub-bids.

The resulting budget may bear only nominal resemblance to the original estimates. He may further re-allocate costs to establish the payment request he intends to give the owner. Some additional adjustments may be made to the reports for cost reporting.

Why the effort?

Forty Years Ago

Forty years ago, I worked with a general contractor who took his lead men and principal subs to a motel for a weekend just prior to the start of a project. Three days, much beer, Coca-Cola and lots of pizzas later, we had the logistics of the job and an explicitly detailed schedule for the project down on paper. Our projects usually went like clockwork, and when the unexpected came up, it was readily solved.

We had better construction documents than you usually see today. If changes were required, they were worked out before they caused delays. The *partnering charter* would have been the butt of a joke, but the *horsetrading* was real. Construction planning and logistics were products of good documents, reasonable horsetrading and good people with sharp pencils.

Among other things, these re-allocations can facilitate the contractor's *front-loading* of his payment requests. They can also be used to confound any future attempts to match represented costs and payment schedules with actual costs.

On some projects a general contractor can, considering the 10% - 15% retention he holds from the subs, realize all his profit by the time the project is only 1/3 constructed.

If the day dawns that the project ends up in a disputes arena, and the owner demands, through the documents-discovery process, the contractor's bid estimate, or tries to use the payment breakdown as the basis for evaluating cost overruns, any relationship between these respective cost allocations can be untraceable and purely coincidental.

THE CONTINGENCY IN YOUR BUDGET

If I suggested that you should add 30 - 35% to the planned cost of your project, you'd probably drop this book with a *thunk*. But you might be surprised at how many construction projects exceed their projected costs by this amount or more.

Not that all the reasons for these cost overruns are design defects. And that's the point. Projects run into unknown site conditions, unexpected environmental requirements, overly ambitious inspectors, arbitrary building departments, strikes, acts of Nature, delayed deliveries. Or the owners may want to provide some room for second thoughts.

17

If you were to discuss the "10% contingency" your architect includes in the "estimated cost" of your project, however, you might hear that this is to cover potential design errors. *Nobody's perfect.*

First, 10% is not a magic number. It's often totally arbitrary. Some straightforward projects don't need 10%. Others, on the cutting edges of design and technology or with likelihood of encountering unknowns could need higher contingencies.

But the contingency in your budget should not be considered a resource primarily to compensate for the architect's errors or omissions.

A well designed, professionally coordinated set of construction documents can be delivered with between 1% to 3% in design-required changes. The standard you should be shooting for with the B/BA is to not exceed your original budget by more than 2%.

Reimbursable Costs

In every standard contract, there is a provision for *Reimbursable Costs.* Inevitably, the article is written so as to allow the loosest possible interpretation by the party proposing the use of the standard forms.

As some of the stories in this book suggest, this is an area through which even the most credible consultant or contractor can make up for some of his failures and oversights.

It behooves an owner to deal scrupulously with reimbursable costs—to establish a definite understanding of what they can and cannot include; and to insist upon regular, detailed reporting of costs, together with backup, with all billings.

THE "WUZZAHS!"

"Wuzzah!—Wuzzah!—Look what I've got for you!"

There are construction managers who produce tons of paper, without either the insight to interpret it, or the competence to apply it effectively. It is this sheer quantity of uselessly accumulated paper that some CMs point to to justify the equipment and the bodies they build into their fees and reimbursable costs.

Then there are attorneys who seduce their clients into feeding the total universe of documents from their projects, plus those produced by all the other parties to a litigation, into the masterpiece of technology they and their favored consultants support. The costs can be spectacular.

The cutting edge of this technology used to be microfiche. Now it's optical scanners and recordable CD-rom disks.

The costs of copying all the available documents on a project can be spectacular, while the benefits can be minimal.

The production of vast amounts of project documentation does not of itself provide any benefits for a project. It takes the insights and competence we've been discussing for it to have value.

Scoping Reimbursable Costs

Project 1: The hospital addition, budgeted for $8 million, was being built in a city 140 miles from the contractor-CM's home office.

The contractor-CM's fee was 4.5%. The contract provided for reimbursable costs in addition. Their definition was the loosely worded one in the standard CM contract the contractor provided. At project completion, the reimbursable costs had added 8.9% to the contractor's fee. Fee and reimbursable costs amounted to 13.4 %.

Only after the owner found it necessary to bring a claim against both the architect and contractor for design deficiencies and the project documents were produced, did it become evident that the CM had been charging as reimbursable costs the round-trip travel of its executives from the home office to the jobsite.

These executives were seldom seen on the jobsite, and then only for a half-hour. They had been traveling to the area regularly to promote other business. The owner recovered these costs in the claim settlement.

Project 2: The C+FF contract contained a provision for reimbursable costs that included "general conditions costs." The Owner's Rep (OR) had estimated that a contractor staff of 8 would be adequate on the $27 million project. The OR had also incorporated into the contract a provision that a schedule of reimbursable costs would be subject to the owner's prior approval.

The contractor proposed a staff of 12. The additional 4 people would have added $485,000 to the general conditions over the project duration. The OR was successful in negotiating a field staff of 9, saving the owner $364,000.

17

It's more constructive and cost effective to structure beforehand a clearly organized and indexed physical system for all the documents that will be produced during design and construction.

The Glove-Compartment Contractor

Mrs. M had operated a day-care center for two years with such success that it was necessary to expand. A building came available in enough time to remodel it before the lease expired on her current space.

She had never undertaken a building renovation before, so she asked around and hired an architect and contractor on the recommendation of people who knew "they did good work." Both the architect and contractor proved to be very "friendly" *(see the first page in chapter 9)* and Mrs. M. was won over.

The drawings by the architect were a little skimpy, but the contractor told her he could "fill in the details" and would do the work for the $200,000 in the budget prepared by the architect. The bank agreed to lend this amount for construction. Mrs. M. took the architect's recommendation that she should use the standard abbreviated AIA A107 form for construction.

The contractor agreed to perform the work in 3 months as shown on a bar chart prepared jointly with Mrs. M. The construction loan required both her signature and the contractor's.

The contractor's first request was for $20,000 up front to finance the job. This was neither in the contract nor part of her loan agreement. Wisely, she refused.

It immediately became apparent that the contractor was not manning the job, either because he was incapable of funding it, or because he was unable to obtain subs. Or both. Work dragged. Some days there were few to no workers on the job. Mrs. M. then learned that the contractor was working on another project. It gradually dawned on her that the contractor's competence would fit into the glove compartment of his pickup truck.

With 4 months already gone, and the job only 1/3 done, Mrs. M, her husband and handy friends started to put in evenings and weekends to make up lost time. The contractor came on the job one Monday morning and complained that she could not do work herself and walked off the job. Then, he started to write letters to Mrs. M. and to the bank about breach of contract and extra work (the architect had overlooked many existing conditions).

The Glove-Compartment Contractor (continued)

Analysis: Mrs. M. did so many things wrong they're difficult to catalog. The first was to accept the unqualified recommendations of others without careful background checks. The second was not to understand the contract she signed. The third was to proceed without sufficient survey work on the existing building. The next was to rely on the budget from an architect who hadn't even checked the existing conditions in the building. She made many other mistakes, which could have led to different scenarios. These are only two of the many possibilities:

Scenario No. 1: Mrs. M. did not keep any independent records of what was going on. She was confronted with a construction claim from the contractor (and his schlock attorney). She had to bring in her own attorney. She had great difficulty getting funding from the bank for the remainder of the work. She had to bond around the contractor's lien. She ended up paying out over $300,000 on her project. Because she lost another 4 months completing the project (the 3-month project became an 8-month project), she lost 40% of her day-care income for the next year.

Scenario No. 2: Mrs. M. at least kept a daily diary (in a bound book) when she visited the job (which was almost daily) in which she recorded the contractor's commitments to her, and the work and workmen on the job. She was in a position to work out an arrangement with the bank under which she changed contractors. She was able to defeat the contractor's claim for wrongful termination, because she had many times notified him that she would have to take over the work if he didn't perform (meeting the 7-day notice requirement in AIA A107). She was required to pay the contractor only for the extra work. All in all, her $200,000 project cost $228,000. Plus a lot of sleepless nights. She lost only about 15% of her clients because her project was completed in 6 months, only 1 month later than the end of her other lease.

The inventory of this physical filing system can be entered periodically into a computerized, effectively coded, flat-file database. This allows specific documents to be accessed quickly when they're needed. The database can be expanded by adding memos, summaries and recaps of events and documents that might have to be referenced later on.

When it becomes necessary to negotiate or resolve specific problems, this computerized access system can be expanded to include detail specific to the issues raised.

The core issues of most projects and of the claims that arise from the problems they produce can usually be analyzed and resolved with just 5% - 15% of the paper that's been generated in the course of its design and construction. Anticipating which documents are likely to make up that 5% - 15% on your project—and setting up the systems to access it efficiently is one of the issues you will want to discuss with your guide.

FOREIGN CULTURES IN YOUR CONSTRUCTION DOCUMENTS?

The design disciplines in the U.S. have acquired—but not necessarily assimilated—a fair-sized contingent of immigrants in the last 25 years. The professional cultures in which these engineers and architects were trained differ in some important particulars from the standards that apply to design in the U.S.

Examples (only two of many possible):

In Europe, design documents are produced with a Bill of Materials, or Quantity Survey. The design consultants produce the quantity take-offs that are, in the U.S., the responsibility of the contractor to develop and extend. In Europe and

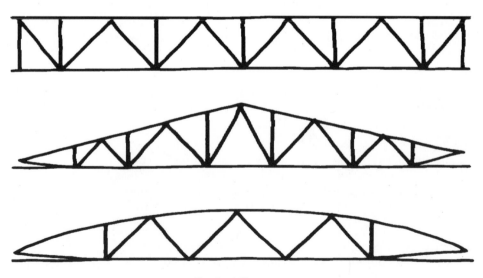

Typical Trusses

other areas, contractors use the quantities that are provided as elements of the construction documents. The owner assumes responsibility for any changes in quantity.

Many construction elements that are delivered as pre-manufactured building components in the U.S. are designed from the bottom up in other countries. In the U.S., the concern

---

Foreign Influences? On My Project?

The building design called for an open-web truss system for the roof. In the U.S., there are several pre-manufactured truss systems that have been pre-approved by local and state building departments. They are so standardized that they can be selected from a catalog. The subcontractor quoted an appropriate system.

The architect and structural engineer rejected the subcontractor's shop drawing submittal. It was resubmitted and re-rejected. It took 4-1/2 months before the contractor could obtain approval of the truss system, when it was approved as originally submitted. By the time the roof trusses could be delivered, it was mid-October. By that time, the site became mired in the worst early rains in memory. The roof could not be completed until early January.

The entire project was exposed through the worst winter storms in 10 years. The contractor requested compensation for the extra work involved in winter protection, disruption, replacement of materials and rusted equipment, and the delay, acceleration and escalation costs he had experienced.

The structural engineer had been trained in a culture 13,000 miles away. A culture in which everything is designed from the ground up. He selected components from catalogs of four different manufacturers, none of which provided separate components.

On a $6 million building, the costs to the contractor exceeded $1.3 million. A protracted lawsuit ensued. The aggregate cost to the parties (paid to attorneys, consultants and courts) was an additional $600,000.

Despite the absurdities inherent in trying to justify the structural engineer's design, the architect's stonewalling, the architect's and CM's failures of responsibility and the owner's obtuseness, 18 months of agonizing hearings and heinous legal costs were expended before the owner was required to pay the contractor the full amount of his claim. Since he had, however, spent a substantial amount of money to reach that settlement, no one won, again, but the attorneys.

17

and competence of architectural offices to check and coordinate their design products has declined. This has led to some costly experiences for owners and contractors. When design offices and their sureties go into Denial Mode, these experiences have had disastrous effects on all but the attorneys' pocketbooks.

COMPUTERS—TIME AND PAYMENTS

On conventional construction projects, it's typical for 7% to 15% of the work activities in an adequate CPM schedule to be "critical."

Computers, however, have made it possible for consultants to produce schedules with as many as 83% (the highest I've recorded to date) of the work activities on the "critical path"—with no logical support for their critical relationships.

This is how it's done:

The Hidden Algorithm

The CPM consultant drew a network diagram of 700 work activities on a $22 million project. The borrowed specifications—which the architect hadn't read—required that all activities have durations of 10 working days or less.

Without having made logical connections for them in the diagram, the consultant—using a feature *buried in the computer program*—produced a computerized schedule in which 581 (83%) of the activities had zero "float"—that is, they were converted into "critical" activities. The architect, never having seen anything but bar charts, accepted the schedule.

That meant that *any delay to any one* of those 581 activities would warrant the contractor claiming an extension to his contract, plus all the attendant impacts to the owner, *plus* the potential damages the contractor could claim for delay and extended overhead costs.

After several months of claimed delays—and with 120 change orders lying on the architect's back desk—the owner brought in someone who demonstrated that only 98 (14%) of the activities were "controlling work." The consultant was able to negotiate the outstanding change orders at 50% of their claimed amounts and to put the project back on a reasonable track.

What computers can do for—or to—schedules, they can also do for budgets, cost reports, payment reports and a whole sheaf of documents that are supposed to provide an owner with confidence that his project is on time, on budget and on program.

There's a natural enough human tendency to believe that if they're neat, if they appear to be organized, and if they're couched in the jargon that surrounds the subject—in our case, construction—then there isn't much to challenge.

But here's another example that puts a lie to this presumption:

> There are specifications in the industry that require that payment-request forms be developed by dollar amounts assigned to all of the activities in the CPM schedule.

> This is a fundamentally good concept *if you have a reference against which to test the allocations for reasonableness.* (We will once we undertake our B/BA analysis.)

> But otherwise, the distribution often provides so much data that little analysis is invested in it. The owner accepts the cost distribution out of hand as a basis for measuring progress.

> Then, monthly, percentages are applied to these dollar amounts to [1] request payment and [2] to demonstrate progress.

BUT—if you divide the hypothetical costs of a hypothetical project into three equal periods and analyze the work in the three periods separately, you might find some startling information in the first period. The contractor could be asking compensation at the rate of $930 per cubic yard for concrete in place. ($225–$280 might be typical).

If you do the same for the equipment and finish work in the third period of the project you might find that equipment and fixtures are allocated half of what they would cost wholesale, with no provision for the costs to install and test them. (See the cashflow figure 14.4 on page 196.)

The contractor expects to put his profit in the bank 1/3 into your project. You have to hope no subs go belly up, and that there's enough motivation—and cash—left in the contractor's account to finish the project.

17

Contractors will do what they can to front-load the costs onto early items in the schedule—to work on the owner's money. It's a practice—considering what we discussed above about estimates and budgets—that can't be entirely eliminated. But an owner should be aware of it, and establish some reasonable limits when evaluating payment allocations.

GIGO

Unfortunately, computers have become so universal (epidemic?) in our culture that it's sometimes difficult to find anyone who can pick up a pencil to cross-foot a column of numbers or manually calculate durations in a CPM schedule.

Even in jobshacks, where you were once out of business if there wasn't a dusty calculator on the back table, there are few tools around with which to check someone else's computer output.

So if the computer was fed bad data, or the computer had software with a bad digestive system (a *bug*), or if the software has a *built-in* facility to mislead—it can go undiscovered for years.

I was once retained to analyze the buy-out of a company that had reported, at the time of its sale, that it was projecting a $2 million profit on 12 ongoing projects.

The sale was consummated. Two years later, it was evident that the records of the company, even at the time it was sold, should have projected a $1 million loss. Obviously, there was major litigation.

There was no fraud. The error was unintentional, although there was some carelessness. The computer systems used to project budgets contained a bug. An algorithm built into the software refused to extend actual unit-cost experience beyond the estimated unit costs—so that projections never reflected what was actually occurring on multiple projects.

If it had not been possible to analyze and manually override and recalculate the relationships contained in the reporting system, the company—which had continued to use the system—might not have discovered the source of the error for a long time.

254

It is possible, wittingly or unwittingly, to bury the potential for so many errors (in both the software algorithms of computers and within the data they ingest), that GIGO (garbage in—garbage out) has become a costly experience for everybody on all sides of the construction industry.

Unless an owner develops the facility and patience to challenge computer data, the experience can become very costly. And the results of disputes resolutions entirely unrealistic.

Computers are excellent tools. I've used them since 1959 with gratitude for the time they've saved and for their facility to select, sort, color-code and present information. But computers can also be insidious when used indiscriminately or with malice aforethought.

Like putting a Maserati under the fanny of an adolescent.

A discussion about the abuses to which computers can be put could fill several chapters. For now, one more caution will have to do.

The Question to Ask

The answer to the question: *"What percentage is completed?"* as reported for the purpose of payment has **no essential relationship to the status of the Contract Work or duration.** In order to obtain a realistic projection of the time remaining on your project, an entirely different double-question needs to be asked: *"Where are we and what do we need to do about it?"*

255

18

AIA DOCUMENTS AND OTHER STANDARD CONTRACTS THAT SHOULDN'T BE

We've discussed many of the concerns and the shortfalls to be found in the standard contracts offered by design consultants, contractors and construction managers.

We've also considered the interest that others have in the performances defined in these agreements—attorneys, the law, insurers and sureties.

We've only touched sparingly on the owner's reciprocal duties as they relate to these agreements. We'll amend that as we get into Parts 8 and 9. Meanwhile though, as an owner, you will find it useful to lodge somewhere in your mind—and probably include in your notebook—the other side of the coin as we consider some of the specific articles in the AIA contracts.

Every performance defined in these contracts relates in some way to something the owner wants, or needs to do, or needs to be able to measure, if the project that all of these contracts are pointing toward is going to be successful.

Those Worn-Out-Tracks

Trying to understand these agreements without assistance is like trying to read a guidebook to a foreign country written in the strange language of that country. The language these agreements contain is often unintelligible to even those who put the forms in front of you for your signature.

Ask any of these proposers of services when they last read all the way through those documents, or when they last explained the respective warranties and liabilities in them to their clients, their staffs or their subconsultants.

Few, if any, of these sincere offerers of Owner Representation advise the owner beforehand, candidly, what his exposures are. Or how to avoid them. Or, especially, how to measure the performance of the offerers themselves.

257

18

The owner is usually out in the middle of his own project land-scape, deep into the middle of the construction contract before what he *didn't* understand hits him.

Even then, and even after a project is propelled into a construc-tion dispute, the owner is in no position to sort out the pieces because the relationships between causes and effects are clouded by everybody's assertions of innocence. And by the intervention of attorneys who speak their own, unintelligible language.

Besides, the attorneys won't let the owner talk about his project anymore, anyway, because it might sink their (and his) prospects.

Most design and construction agreements require all the strang-ers the owner invites into his *new* landscape that follow tracks that were laid out *30, 50 or more years ago.*

Those tracks are the administrative procedures in the agreements and specifications developed by the American Institute of Archi-tects (AIA), the Associated General Contractors of America (AGC), the National Society of Professional Engineers (NSPE) and similar industry organizations—organizations that are motivated prima-rily by the proprietary concerns of their respective members.

Although each of these organizations has different memberships, and therefore special concerns for the liabilities of its members, each of the other strangers operating under these agreements and specifications has a different viewpoint about where those tracks should lead.

Many contractors are disdainful of the conditions in the AIA stan-dard forms. They wonder, with good reason, why the architect should be indemnified against his own failures in performance. (As we've discussed, it's because there's pressure from insurers to keep it that way.)

Every stranger that comes onto the owner's project has a differ-ent reading of those tracks, and a different compass setting. It's almost inevitable that those different compass settings will lead to unexpected meetings at unplanned intersections: accidents waiting to happen.

Since, when the accident happens, it will be on the owner's land-scape, it's a sure thing the owner will have some surprise bills for the damages.

At that point, the owner will be hard-pressed to count the new tracks of the strangers who will appear in his landscape. Lawyers, claims consultants, sureties, arbitrators, mediators and uninvited camp followers will attach themselves for whatever nourishment they can extract from the owner's resources.

And along with all the legal costs this kind of accident produces, the owner will experience costs from delayed occupancy, deferred income, increased administration, increased carrying costs, and significant disruption to his other activities.

Whether the tracks laid down in the standard AIA agreements (for either design or construction) are aligned with the owner's own compass and directions has been a source of discussion in the industry for a long time.

But if there was a question before, evidence has accumulated in the past 30 years—and particularly in the last 15 years—that demonstrates how far the compasses of these associations, their members and their sureties have swung away from the owner's lodestar.

Bless 'Em

When did an owner last get down-and-dirty with the architect's, the construction manager's or the contractor's agreement and challenge what those documents promise, and how those promises would be measured, before signing them? When did an owner last compare the administrative duties in one of those contracts with the others?

Some owners ask their attorneys to "look over" the contracts. The attorney tinkers with the disputes-resolution clauses. He may review the insurance requirements. He may even discuss the contract amount. That done, he blesses them with his particular holy-water and the owner executes them with high expectations.

But if an owner compares these agreements with their equivalent documents as issued 30 years ago, he will find that the administrative requirements of those agreements have been modified more to provide protection for the firms that offer them than to benefit the owner. And not at all to keep pace with technological developments.

18

These canned agreements and specifications, instead of being thoroughly redefined to secure the interests of the owner, have been supplemented with more, equally ambiguous documents.

These add-on, often redundant, documents define new panaceas to cure the problems created by the original documents. They include *comprehensive services, construction management, segregated-prime contracting, total quality management, partnering, fast-track construction, flash-track construction,* ad infinitum, ad nauseam.

They're like government regulations: no one understands them, and while you're trying to comply, they're subject to continuous reinterpretation. More and more, they're being interpreted in the courts, in mediation and in other ADR venues against all the acceptable rules of fairness and common sense.

The surprise implied by the owner's inner voice: ***"Where are we and how did we get here?"*** is a surprise only to the owner.

It's in the nature of construction that the owner will be presented with surprises. But the last place they should be hiding is in his Contracts or in the Construction Documents.

Although there's a broad spectrum of standard service agreements in the construction industry, the content of the AIA documents provides a preliminary introduction and road map for consideration of the provisions you will want to modify, or replace, in the contracts on your project.

Once through these chapters, and then Parts 8 and 9, you will probably want to plan several meetings with both your attorney and your guide.

If yours is a public agency, you will find that the contracts and specifications your agency now uses contain many of the same shortcomings of these agreement forms. That's because most of them have been translated by attorneys who, whatever their relationships to your organization, were as ignorant as you were of what occurs on a project through the processes of design and construction.

The only view many of these attorneys have had—including many corporate in-house counsel—has been from behind the windows of a real estate office, in trials and negotiations of personnel and financial matters, or through receipt of the construction claim that just came into your office. Many never got closer to a construction dispute than through the course they took in Contract Law.

It's useful to emphasize once more how important the insights of the guide we will be meeting in chapter 22, and of the attorney the owner elects to work with, will be to help the owner develop a working understanding of what these documents mean and how they can be negotiated and implemented.

> It's strange to someone, who's been in the industry for over 40 years, that most owners turn to these professionals only *after* their projects get into trouble. It might be a good time, once you've been through this chapter, and if you haven't done it yet, to set up some meetings with your guide and your attorney before you move onto the finer points of establishing your Project Plan.

As you read through the following excerpts, you should know that the AIA has prepared a parallel system of commentaries intended for the use of its members as they proceed to resolve agreements with clients. These commentaries are specifically directed to the design professionals in terms of [1] Practice Pointers and [2] Liability Alerts.

The Design Agreement: B141 (the 1987 version)

Article 1: *The Architect's services shall be performed as expeditiously as is consistent, etc., etc....*

The first article in B141 contains the first debatable word in the architect's design agreement. What does *"expeditiously"* mean? AIA commentaries suggest that the statement that "time is of the essence" *can increase the architect's exposure to liability*. It has been consciously deleted from the standard design agreement.

Nothing is said, either in the agreement or in the commentaries, about the owner's exposure if the documents are completed unseasonably late, missing either good construction weather or an ideal bidding period.

18

The next sentence in Article 1 reads:

Upon request of the Owner the Architect shall submit for the Owner's approval a schedule for the performance of the Architect's services.

It's in the owner's interest to require this schedule and to be very specific about the milestones for measurement and payment of design documents. But the AIA commentaries suggest that the architect make schedule promises only about "things" he controls.

You are now on notice that the architect will not volunteer anything the owner doesn't require him to do.

You are also on notice that you will need to get your own act together, since you can be held accountable if *you* delay the design process.

Right from the get-go to the end of the project you will have to take the initiative.

If you have invested in your Plan, this could be the first test of the Architect's skills. How does he view schedules? What does he know about definitive ways to schedule (a logically sequenced critical-path schedule that coordinates all disciplines)?

How prepared is he to subject his own production to that type of scheduling? If not at all, what alternative does he offer to assure you that your project will meet its deadlines? What does he know about construction budgeting? Is he prepared to put his document production on a measurable basis for payment?

Articles 2.2 through 2.6 define the Architect's basic services.

Article 2.2, Schematic Design Phase goes on to say: *The Architect shall review the program furnished by the owner* and that they shall arrive at *mutual understanding.*

It further says that: *The Architect shall provide a preliminary evaluation of the Owner's program, schedule and construction budget, each in terms of the other.*

This is where your preliminary Program and Plan come into the picture, to keep the architect's *chaos* out of it.

Article 2.2.4 commits that: *The Architect shall prepare... Schematic Design Documents consisting of drawings and other documents illustrating the scale and relationship of the Project components.* At the end of this phase he will *submit to the Owner a preliminary estimate of Construction Cost based upon current area, volume or other unit costs.*

You will need to establish your own definition and description of the documents you will require to confirm that schematic design is proceeding in relation to your Program and budgetary requirements. There are no definitions in the agreement apart from what you read in article 2.2.4.

No specifications are mentioned. Although specifications would not be very far advanced at the end of the schematics phase, it could still be very important that their direction is clearly outlined so they can be tested against your Program. Something more specific than "area, volume or other unit costs" will be required to test their direction in terms of your budget. (At the least, descriptions of the structural systems, building enclosures, quality of finishes, types of equipment, etc.)

Article 2.3, Design Development Phase commits the Architect *to prepare for approval of the Owner Design Development Documents consisting of drawings and other documents to fix and describe the size and character of the Project as to architectural, structural, mechanical and electrical systems, materials and such other elements as may be appropriate.*

Article 2.3.2: *The Architect shall advise the Owner of any adjustments to the preliminary estimate of Construction Cost.*

There is no requirement for the architect or other design consultants to produce specifications in this phase. The first time specifications are listed is in **Article 2.4, Construction Documents (CDs).**

The question immediately arises: *how reliable are the estimates of construction cost the architect is supposed to provide the owner at the end of each phase? Or the construction duration contained in the CDs?*

The AIA commentaries suggest that the architect and the client need to discuss realistically the relationships of scope, quality, price and time. If the owner doesn't take the initiative, though, it's not a requirement in the agreement. If the architect doesn't take this suggestion seriously early in the design process, the owner can find himself well past the point of no return before he knows it.

The AIA commentaries, directed to the architect, also point out that the owner is obliged to act reasonably and in good faith in approving design documents and making adjustments.

That is good law and absolutely true.

But the corollary as it relates to the architect is not included in the commentaries. The standards the architect is required to meet in coordinating the work of his subconsultants is left in the limbo of *what another professional acting reasonably would do.*

The owner needs to require that the architect be held to a standard equal at least to the warranty that the architect's documents will be making—on the owner's behalf—to contractors.

There is nothing said in the AIA agreement about how the architect will accomplish this.

It provides no standard for his design schedule. It provides no basis other than "75%" or its equivalent for measurement of progress. It does not correlate either of these to payment for his services.

It does not hold the architect to a standard for compliance with the owner's Program, nor to his budget. It does nothing to assure that design will be completed within a reasonable time, as measured by optimum weather for construction, nor for optimizing beneficial bidding seasons.

It says nothing about the architect's accountability, nor about what the E&O policy *really* means as protection for the owner against design defects.

The architect becomes, once the design contract is signed, *the sole arbiter of his own performance,* which the owner can only evaluate in the context of *what another professional acting reasonably would do.*

To recap: there is no provision in the AIA B141 design agreement for independent evaluation of [1] design coordination, [2] buildability, [3] consistency with the owner's program, [4] the design team's production schedule, [5] the assessment of design product versus payment, [6] the projected construction duration, [7] the projected construction cost, or [8] the adequacy of the administrative procedures to be implemented during construction, among other areas of the architect's performance for which measurement isn't covered.

Article 2 is important for the owner to annotate in its entirely because the phrase *assist the Owner* is used repeatedly throughout.

If the owner doesn't have a good, clear sense of direction, and very specific criteria for where he expects to go with his project, its schedule and its budget... well, there's no need to repeat where that leads.

Construction

Article 2.6.2 provides that: *The Architect shall provide administration of the Contract for Construction as set forth below and in the edition of the AIA Document A201... unless otherwise provided in this Agreement.*

We will discuss in chapter 19 the options the owner has with regard to administration of construction.

Many of these options will take the architect out of the role of construction-contract administrator.

But he will, no matter the form of construction contract, be critical to the successful completion of the project. He will remain responsible for inspections, design compliance, review of submittals, RFIs, payments and sign-off of final completion. He will be pivotal to issues of design defects and changes and will prepare the design components, at the least, for all changes.

If you were to go through the AIA agreement that integrates a construction manager into the project (AIA B801), or the AIA Design/Build document (AIA B901), you will find repeated use of phrases that require the construction manager or the contrac-

18

tor to *assist the Architect... provide recommendations... provide for the Architect's review and acceptance of a project schedule... advise the Architect... consult with the Architect.*

What if the Architect doesn't accept? What is his obligation with regard to recommendations? What are to be the results of these consultations? Who will resolve disagreements, or recalcitrant or delayed performances, or recognition of responsibility?

It's the lack of answers to these questions that makes this introduction to the design agreement, limited as it is, elemental to the owner's concerns for the success of his project.

Any owner concerned about the respective responsibilities of the parties he retains has a lot of work to do to nail down schedules, budgets, progressive design documents, and milestones, duties and responsibilities of the parties.

The relevance of the owner's comprehensive Project Plan becomes progressively more apparent as you read deeper into these agreements.

The General Conditions of the Contract for Construction, AIA A201, 1987

The AIA commentary associated with AIA A201, the construction contract for a lump sum, notes that: *the general conditions are not completely applicable without adaptation and fine tuning.*

Amen.

The fine-tuning needs to be the owner's, not the architect's.

In the standard General Conditions:

> **Article 1** identifies the documents that make up the Construction Contract. These are the documents that spell out all the terms and conditions of the agreement between the owner and the general contractor.

> **Article 1** specifically excludes the bid documents.

> Why? Because the AIA commentaries assert that the bid documents may contain provisions that *conflict* with the contract documents.

They had better not.

It's the responsibility of everyone developing and contributing to the bid documents to see that they don't. This necessarily includes the owner, since some of the terms of the bid documents need to be determined by the owner.

Too many construction disputes have been raised because of discrepancies between the construction documents and the bid documents. The bid documents have sometimes prevailed even when they're not identified as Contract Documents. Using their possible inconsistency as a reason to keep them out of the Contract Documents is bad advice—even if, in the final analysis, they are not to be considered Contract Documents.

The commentaries also make the point that oral discussions during negotiations have no contract effect unless reduced to writing agreed to by the contracting parties.

That's good law, too. It points up the fact that the owner can't demand anything that's not in the contract when it's signed.

But it overlooks the fact that oral commitments made by the owner's representatives in pre-bid conferences (as one example) can come back to bite the project even if they're not confirmed in writing.

Skipping ahead:

> **Article 1.2.3** says: *The Contract Documents are complementary.* The commentaries go on to say that the contract documents should have no order of precedence, so it is the architect's responsibility to interpret the documents and performance under the documents.

This is a bad precedent. This is the beginning of the absurdity that the architect will determine the rightness or wrongness of the performance of the parties to the contract—the owner and contractor—under the terms of the contract.

The precedence of the documents refers to which document will prevail if the information in them is conflicting. It must be defined in the Contract.

> **Article 3.2.2** specifies that: ***Errors, inconsistencies or omissions*** *discovered (by the contractor) shall be reported to the Architect at once.*

18

When the articles below are considered in this context, it becomes clear that the owner has a lot of work to do to make the AIA A201 *his* document and not the architect's.

Article 4.2.11: *The Architect will interpret and decide matters concerning performance under and requirements of the Contract Documents on written request of either the Owner or Contractor.*

The commentaries make it clear that *performance* is not limited to the technical requirements of the drawings and specifications but also includes *judging (the owner's and the contractor's) compliance with the terms of the contract.*

Article 4.2.12 says that: *When making such interpretations and decisions, the Architect will endeavor to secure faithful performance by both Owner and Contractor, will not show partiality to either and will not be liable for results of interpretations of decisions so rendered (by the Architect) in good faith.*

And

Article 4.2.2 provides that: *The Architect will visit the site at intervals appropriate to the stage of construction to become generally familiar with the progress and quality of the completed Work.*

Even more specifically than in some other articles, it's made clear that the Architect *will not be required to make exhaustive or continuous on-site inspection and/or will not be liable....*

Somewhere along the line the AIA has been able to divert owners' (but not contractors') attention away from the fact that the most frequent and significant causes of disputes between those two parties are over the information the architect has put into the contract documents.

One more article of interest in the AIA general conditions for construction, before we add up the score.

Article 4.2.7: The Architect's action for review and approval of the Contractor's submittals *will be taken with such reasonable promptness as to cause no delay in the Work or in the activities of the Owner, Contractor or separate contractors,*

> *while allowing sufficient time in the Architect's professional*
> *judgment to permit adequate review.*

Sounds good but is entirely hollow.

It calls for the Architect to have a good handle on what's going on with the job. (How does he do this with site visits at *appropriate intervals?*)

It also requires him to be competent in reading and analyzing detailed progress schedules.

> There are neither definitions nor provisions in the standard AIA A201 regarding the level of scheduling, and certainly none in any of the contracts that commit the architect to the competence required.

In the environment of construction claims, architects (their attorneys and their sureties) have argued that *"reasonable promptness,"* can be as long as 8, 10 and even 16 weeks after the architect has received submittals. These positions have been asserted even when it has been conclusively demonstrated that design errors were at the root of the asserted disputes. And when reading of the CPM schedule demonstrates that these periods of time exceeded the "float" on the work that was delayed.

What goes for submittals holds equally for RFIs, change proposals and similar transmittals of information between the contractor and architect.

Despite all these discrepancies and inconsistencies, many architects, their sureties and their attorneys have tried to transfer the impacts of delays, and even the costs for correcting defective design to contractors through a misapplication of Article 3.2.2.

The extraordinary thing is that this assertion has prevailed with arbitrators and mediators, and even with some courts.

> In 1968, in an effort to make the AIA documents we've just gone through responsive to the requirements of a client with a continuing building program, I attempted a side-by-side, article-by-article modification of the AIA A201 general conditions. I abandoned the effort when, before I got through Article 6, I had overwritten more than 65% of the text.

18

Working with the client's attorney, our firm developed a new set of conditions that contributed to the successful design and construction of 20 projects with consistently minimal design changes and without claims. These new contracts also received constructive responses from contractors.

Other Sources, Add-Ons

The AGC, the NSPE and other associations have generated their own contract forms. They generally follow the outline of the AIA documents. Makes sense: the same phases and functions need to be performed on all projects, whatever their types, size and design requirements. And as you remember, the AIA was there first, so it established the outline.

The tilt in these proprietary contracts benefits the members of the association that developed them. But the phraseology can be sufficiently subtle, and contain so many parallels, that it takes close study to determine what assurances, what loopholes and what promises each contains.

In the descriptions of work scopes and responsibilities there can be considerable redundancy without delineating distinctive responsibilities or adding value. They frequently facilitate loopholes because of their ambiguities.

For many years, a joint panel has been working to integrate the elements of the AIA documents and the documents of several engineering associations, represented by the Engineers Joint Contract Documents Committee.

The effort has value, but the end produced (if and when it becomes generally accepted and applied) will require the same scrutiny the AIA documents deserve. Whether or not the effort results in a universally accepted document, it will have been developed by associations that along with their sureties, have vested interests in minimizing the liabilities of their members, at the possible expense of some other party.

The same concerns regarding liabilities—the owners' and the professionals'—are discussed at length in the commentaries that discuss the content of these agreements.

An Example from the AIA Commentaries

The term *"probable construction cost"* used in earlier documents was changed in the 1987 version of the AIA B141 to read *"preliminary estimate of construction cost."*

The commentaries note that the change was not intended to change the substance of the term to mean a *detailed estimate* (a term used elsewhere in the agreement) and that the client (owner) needs to understand the *tenuous* and *tentative* nature of the *preliminary estimate.*

The commentaries specifically note that the client is not encouraged *or permitted* to rely upon the preliminary cost estimate (provided by the architect) for *more than the design professional intends.*

The estimate the architect commits to deliver with the final construction documents is specifically identified as a *preliminary, tenuous and tentative estimate of construction cost... on which the owner is neither encouraged nor permitted to rely.* This leaves in limbo (a large area of potential pain that is without defined boundaries) what the owner is supposed to rely upon for his budget.

What does the design professional intend? That's an issue the owner needs to establish *before* negotiating a contract, or even issuing an RFQ for architectural services. The design agreement also will bind the architect's subconsultants, so their agreement needs to be brought into context, too. That is why the owner needs to think these things out in advance and to require that all the principal design subconsultants be made party to all pre-agreement discussions.

The owner always has—and *BIT* recommends—the option to bring in his own, independent resource to develop a detailed construction-cost estimate. He can incorporate terms into his design agreement that say that discrepancies in the documents, which affect his project budget, be revised at no cost before construction. But this requirement, as we suggested above, has to be in the original contract. Which is what the Plan and the B/BA are all about.

The practical way to assure budget compliance is to check it progressively, in progressively expanded detail, as design ad-

18

vances. All of which will require careful resolution with the architect before committing to the design contract. As will a myriad of other performance requirements.

The Tips of the Bergs

In this brief overview, we've looked only at the tip of an iceberg with potentially dangerous subsurface implications for an owner's project. The articles we've reviewed, and considerably more, need to be carefully redrafted to be responsive to an owner's need to realize a successful project.

In the AIA A201 for construction, alone, there are some 25-40 paragraphs in which the requirements for measurement and documentation of the architect's and contractor's performances during construction need to be spelled out in detail.

Responses to submittals, RFIs, changes, etc. need to be specifically related to a sufficiently detailed construction schedule. The construction schedule (whatever format is identified by the owner) needs to be maintained concurrent with progress by the contractor to reflect accurately and concurrently the actual construction- and design-administration progress on the project.

> This is a particularly significant area that can result in considerable arm-wrestling with the contractor.

> One particularly effective method is for the owner to maintain parallel tracking of progress in the same scheduling system the contractor has been required to use.

> With the owner receiving comprehensive documentation concurrently with progress, this puts the contractor on notice, at the least, that he can't sneak up on the owner with cries about someone else's performance if his own is deficient. (See comments about sharing viewpoints of *time* in chapter 14.)

Meeting the standard of producing buildable construction documents requires the architect to manage his subconsultants and the interactivity between design and construction within very specific limits—limits defined by the approved construction schedule. That requires well-developed facility with, and commitment to, the use of tools for measuring and documenting performances, time and money, and for timely responses to possible unexpected conditions.

These are not attributes for which architects are noted.

And yet the AIA documents encourage—in fact, unless they're modified, the documents mandate under Article 2.6.2, above (unless the owner specifically deletes this scope)—that the architect shall manage the design and construction processes. The documents contain no descriptions of how either of those processes will be measured or documented.

Then, the documents go on to promote the architect's management of construction compliance while leaving unspecified any accountability for the architect's own performance during construction.

Even when a "construction manager" is introduced into the mix through one of the add-on AIA forms, the agreements are ambivalent and ambiguous about how changes, documentation, payment, and design services will be administered—or how differences and decisions will be resolved.

AIA A201/CM

This alternative agreement to the A201 brings another party into the mix:

> **Article 2.3.2:** The Architect and Construction Manager will advise and consult with the Owner. (They) will have authority to act on behalf of the Owner only (as modified) in 2.3.22, which paragraph says: *The duties, responsibilities and authority of the Architect and Construction Manager as the Owner's representatives during construction will not be modified or extended without written consent of the Owner, the Contractor, the Architect and the Construction Manager....*

What consensus will there be among these four parties if things go badly? Who's the strong horse, seasoned in confrontation?

And still, even with a construction manager in the mix, **Article 2.3.12** in this A201/CM contract for construction, says:

> *Claims, disputes and other matters in question between the Contractor and the Owner relating to the execution or progress of the Work or the interpretation of the Contract Documents* (all of the Contract Documents, not just the design documents) *shall be referred initially to the Architect for decision. After consultation with the Construction Manager the Architect will render a decision in writing within a reasonable time.*

Who's in charge here?

18

Not To Worry—Your Organization Has Written Its Own Contracts?

These inadequacies telegraph themselves right on through the agreements owners translate onto their own letterheads. It requires only laying up one of the forms that counties, cities, private corporations and special districts have had their legal departments prepare, alongside the standard AIA forms to find the same slippery slopes and shortfalls discussed above.

The Owner's Options

It's time to look at some (only some) of the optional contract formats available to an owner. The options can substantially change the owner's relationships to his architect and his contractor and to the way things will be done on his project.

Some of these have real merit. Some of them are just the reflections of the absurdities and panaceas we discussed in chapters 16 and 17.

All of them demand that the owner carefully evaluate the resources he can bring to the project in relation to documenting and measuring the performances these other contracts promise.

Design and Construction Agreements—What To Do?

We've started to look at them closely. How can you improve them?

First, take a note from the salmon. There's no point spawning a new system of contracts just to die at the end of that effort, before you even begin the journey for which they're intended.

As archaic as the AIA and similar contracts and specifications are, they have set a format and an outline that parallels the processes we visited in chapter 7, MAPPING THE TERRITORIES. You need to follow that outline or those strangers out there will turn their backs on you.

What To Do? (continued)

Using this outline (let's say you start with AIA B141), review the shortcomings in the design agreement as described in chapter 5 (where contractors look for potential design defects), and chapters 8 and 9 (the design team's scope of work during design).

Using your notebook, outline where you would improve the architect's performance and your ability to measure it. (WHERE HAVE WE BEEN? is organized to help in this process.)

You may have to consider phased introduction of some enhancements over time, if your organization is hidebound (chapter 25 comments on this issue).

Insert the requirement for the Buildability/Budget Analysis (chapters 23 and 24) into this scenario. The place to start this insertion is under article 2 of AIA B141.

The end product will be an outline of the improvements you have decided you want to make, article by article, so that you and those who represent you will have the tools with which to measure the products of design under the B/BA. This outline won't be final. It's a point of departure from which to start your discussions with your guide.

Since we're discussing a legal document, one that will set the content and direction of your Design Contract, you will need to bring your attorney into this discussion early, too.

When you get together the three of you will also need to consider risk management.

If you're going to require responsive performances from your design team, and if you believe that you will not be able to obtain comprehensive coverage for design errors and omissions, you might have to discuss going "bare" (without insurance coverage) or other strategies for setting limits that you will put on the architect's margin for error when the construction contract is confirmed.

You might consider obtaining litigation insurance against the potential for claims. Since much coverage contains a "wasting" clause, you may be better off with some of these alternative strategies in any case.

You will have to go through this process again with the construction agreement and general and special conditions. You will have most of the design period to accomplish this, unless you intend to proceed with a design/build contract (see chapter 19).

Blind Man's Bluff

Your architect provided you with a "preliminary estimate of construction cost" (1987 AIA B141) of $620,000. You relied on it as your budget.

Four contractors told you your project would cost between $740,000 and $763,000.

The backup for your architect's budget consisted of 18 line items. (Typical: "Wall systems, $67,500.")

The contractors' bids were based upon quantity take-offs of the labor and materials respective trades estimated for: "concrete," "structural steel," "metal studs," "drywall," "windows," "insulation," "painting," "hardware," "piping," "HVAC equipment," etc.

Against what criteria do you start to negotiate realistic reductions or "value engineering" with the contractor? Or are you dependent upon the mercy of the contractor to *tell you* where he can reduce costs?

19

ENDLESS POSSIBILITIES FOR DESIGN AND CONSTRUCTION CONTRACTS

Contracts: The Options

There really are almost endless options for the relationships that can be established between the owner and the contractors on a building contract. For each basic format—lump-sum, cost-plus, Design-Build, segregated-prime—there are numberless considerations on how costs and payment will be measured, time scheduled, supporting documents provided, disputes resolved and scores of other considerations, any one of which can have great significance as to how the parties deal with one another.

Whatever the promises, there never was—and is never likely to be—a contract that was so unequivocally written that it cannot be challenged.

We will survey only the broad scopes of them. Nothing could be more important than a careful reading of every contract the owner signs for the development of his building project.

It becomes apparent, when you sit through the testimony of participants to a construction dispute, that few of them have real understanding of—or have even bothered to read through—the contracts that defined their duties.

When we consider the options available to the owner of a building project, we're confronted with more of the ironies of the construction industry:

- Few understand what they have promised, or what has been promised to them.

- Whatever the promises, there never was—and is never likely to be—a contract that was so unequivocally written that it cannot be challenged.

- Even the most carefully written, apparently unequivocal contract can still become the victim of the law's whimsies.

19

That is why the only quotient that people can rely upon when they sit down to exchange promises is mutual respect based upon knowledgeable candor and honesty.

If one or more of the parties to a contract doesn't maintain those virtues, the next best resource on which the other party can rely is comprehensive documentation based upon competent understanding of the promises exchanged.

A Natural-Enough Tension

There is an inherent tension between the design and the construction agreements. It is contained in the terms of those agreements, and in the performances and motivations of the parties.

Any modification to the format or the content of a construction contract will have implications for its related design contract. They cannot be considered separately, or in isolation.

Confrontational? Never...

The AIA and AGC have a newfound relationship. The organizations have cooperated to produce the 1997 versions of the standard agreements (B141, A201, etc.).

No sooner was it published than the AGC found a discrepancy in the A201 (governs construction) that it had endorsed and the B141 (for architectural services).

The discrepancy has a major impact on the owner's duties—what else?—as they relate to the owner's establishment of design criteria.

Caveat emptor, qui ignorare non-debuit... etc., etc. (chapter 5).

The owner will hear, principally from the design side, that there should be no reason for confrontation between the design professionals and the contractor. This comment almost always comes from the architect and is driven by his potential anxiety about what the contractor's response will be when he picks up the plans.

One of the purposes and the benefits of the owner's Project Plan should be the reduction of this anxiety. As we suggested in the section on *partnering*, the owner can substantially reduce this level of anxiety with an effectively implemented Buildability/ Budget Analysis.

The Lump-Sum Contract

This is the one to which most owners of homes, commercial buildings and public projects are introduced by their architects. It is represented in the standard AIA A201 contract for construction.

In forms modified to meet their legislated requirements and charters, it is the contract format public agencies use.

Whether negotiated or competitively bid, it is designed to obtain from a general contractor his best judgment of what it will cost to deliver *exactly* what is in the construction documents. *Exactly* is the operative word.

For all that the architect protests:

> *We knew it wasn't complete... we expected that the contractor would fill in the details. The contractor was required to tell us if there were any problems with the drawings... nobody's perfect.*

The bottom line is that the owner cannot require the contractor to build what is not in the drawings, or *what cannot be built from the drawings,* without compensating him for extra work.

The contractor can be required only to build what is contained in those lump-sum construction documents unless he's compensated for whatever may be extra, or changed. (Note the owner warranties in chapter 5.)

The contractor can be required only to build what is contained in those lump-sum construction documents unless he's compensated for whatever may be extra, or changed. (Note the owner warranties in chapter 5.)

The Time and Materials Contract (T&M)

This is the second-most hazardous and administratively demanding contract the owner can authorize.

The contractor will provide time records of the labor spent on the work. He will add the cost of materials, together with an agreed markup for the overhead associated with procuring these materials. Overall, he will add a markup for his operations (field and office overhead). And over all this, he will add a margin for profit.

19

There are likely to be no change orders, unless the original scope of work has been specifically defined. The contractor will just perform the work and bill for it. That leaves a barn door wide open for the contractor to just perform the work, bill the owner whatever costs he has accumulated, and for the less-than-scrupulous to pad the bills in the process.

The owner has the responsibility under this type of contract to define limits for the scope of work and for the budget.

Then, he must have competent people following every move made by the contractor. He must have clearly defined procedures for reporting of labor expended. Labor must be reported, independently recorded, and measured by the owner's representative *every working day*. Each day's timecards must be signed off by the owner each day *in order for payment to be honored,* and the owner must have some credible basis with which to measure the reasonableness of progressively expended costs in order to assure he is not being overcharged. It is very easy for a contractor to run unnecessary labor into the workplace to pad his invoices.

There is only one justification for T&M work: that the scope of work cannot be determined in any other way than exploring and exposing the work to be performed.

To the extent that this is a condition that turns up under an ongoing contract, the authorization of T&M work must be carefully spelled out in a scope of work, severely circumscribed by a definitive budget, and closely monitored on a continuing basis.

A T&M change order on an ongoing lump-sum contract can otherwise be an invitation for the contractor to get extra payment for lump-sum work.

> When you think about this a moment in terms of the AIA B141 contract and the architect's duty to make "periodic observations" of the work—the mix of architectural administration of the construction and T&M work would be a marriage made in hell for the owner.

Segregated-Prime (sometimes *Multiple-Prime*)

This is a format under which the owner, or some representative of the owner assumes many of the responsibilities usually performed by a general contractor: planning of construction logistics, scheduling, coordination of trades, organization of procurement, meetings, design administration during construction, change orders, and other essential paperwork.

The representative of the owner that assumes this role is often a stand-alone CM firm (not a general contractor) that has aggressively marketed the idea as a way for the owner to save general-contractor markups.

The owner executes prime contracts with separate subcontractors.

Often, the first *separate contractor*—the one awarded the work for foundation and frame development—is also directed by his owner to *coordinate* the work of the other *separate* contractors. They in turn are required by their contracts to *cooperate* with the first contractor.

They have no contractual relationships. They have contracted only with the owner, who is in turn represented by its own personnel, or by a construction manager who has assumed those responsibilities suggested in the first paragraph above.

So You're Considering Multiple-Prime Contracting?

In New York State, a multiple-prime law (the "Wicks Law"—1912) mandated multiple prime contracting on public projects exceeding $50,000 in estimated costs. It was supposed to save the excessive overhead costs associated with general-contracting services.

An independent study conducted on 160 projects built in New York between 1980 and 1991 demonstrated the following:

- Multiple-Prime projects cost 13% more, and took 60% longer to build than comparable projects not built under multiple-prime.

- On average, multiple-prime contracts took 15.6 months longer to build than similar projects not built under multiple-prime.

- Multiple-prime projects cost $14 per s.f. more than projects of the same type, size, time frame and complexity. The owner's internal costs for monitoring these projects contributed $7 per s.f. of these costs.

19

I previously commented that T&M is the *second most hazardous* contract an owner can undertake.

A segregated-prime contract is *Hazard Maximus*.

In at least a half-dozen workshops with construction attorneys I've asked the innocent question:

"What do you think of segregated-prime contracts?"

The invariable answer from the assembled attorneys has been:

"We love 'em. They keep us very busy."

And yet, believe it or not, segregated-prime contracts are mandated by many states on their public works. Somebody in the dim dark (repeat: *dim dark)* past convinced their legislatures that there would be considerable savings in the markups of general contractors on publicly bid work.

Nevertheless, there are hapless groups of our citizens who have been, and continue to be, seduced into believing that their projects will benefit from this abortive way to save money. Notable amoung these citizens are those who sit on the country's separate and isolated—and thereby defenseless—school districts.

An owner needs to ask:

- What competence, what responsibility and what authority does the segregated-prime CM have to ensure that the documents are unambiguous and buildable?

- When the CM comes onto the project without prior relationships or established credibility with the various subcontractors who will bid the work, what influence and what leverage will he be able to exercise to ensure their performances?

- How does the fee that will be paid to the CM compare with a general contractor's competitively bid fee? What are the actual savings?

- Then, consider the costs to the owner of administering the separate contracts, bonds, payments, change orders, procurement and other paperwork flowing from some 10 or 20 separate subcontractors and 50 - 80 suppliers.

- How many separate disputes and claims might evolve out of these separate contractual relationships? What will be the medium for resolving respective responsibilities? (See the definition of "ripple" in the glossary.)

- How much liability is the CM prepared to pick up in relation to "coordination" and "scheduling" of all the separate contracts, along with the scores of suppliers whose deliveries and performances are as important to the project's success as are the performances of the "cooperating" subs?

To those who've been there in trials or who have considered the histories of segregated-prime contracts, they're mostly time bombs waiting to go off.

Design-Build

The first hut was constructed by a designer/builder. As we came out of the dark ages it was the *master builder* who both designed and constructed Chartres and Rouen. Bramante, Michelangelo, Brunelleschi not only designed but directed the construction and worked the marble of their Renaissance masterpieces.

Design-Build is not a new idea. It was shoved aside as the Industrial Revolution introduced the need for integrating new disciplines. The AIA formulated its first prototype contracts in 1888 in cooperation with the American Builders Association. The architect assumed the role of coordinating these separate disciplines, and left construction responsibilities to contractors.

Design-Build had something of a resurgence in World War II but was temporarily absorbed into a new idea: Construction Management.

Design-Build, promoted in the last few years by the contractors, has experienced new life. Not surprisingly, the AIA considered D-B an intrusion into the territory it had staked out but more recently has bowed to the inevitable and now is trying to find its place in the Design-Build spectrum.

There are even some AIA firms that would turn the idea upside-down again, proposing Design-Build with the architect in a dominant role. It all leads you to wonder how many reversals you need to go through before you end up right where you started.

19

A thousand years of evolution and convolution has put D-B in the same league with other panaceas. The owner needs to consider carefully what contract format is in the best interests of his project and his organization.

The benefits claimed for D-B include that the design as well as the construction will be under the direction of a knowledgeable contractor from the beginning. The contractor assumes the responsibility for assuring that it will be buildable.

It is important under such a contract, nevertheless, for the owner to have a well-defined set of criteria to ensure that the owner's requirements for quality, budget, functionality and aesthetics are also represented in the completed project.

Design-Build can have a natural synergy with construct-lease-back programs, whereby the organization—which today often includes a developer or investor group for the financing—builds and owns the facility, and leases it back to the owner. Since many of these private developers are contractors, or are closely associated with contractors, the contract fits well into build/leaseback.

Because Design-Build has seen a resurgence in the last few years, it has recently been subjected to a number of studies. One of these studies, by the Construction Industry Institute, was reported in the *ENR (Engineering News Record)* of November 1997.

This study found some time savings for D-B, some slightly higher costs, and only a slight edge in quality improvement. This is not the final word on the subject (350 projects were studied in the categories of CM, D-B and a variation on the theme, design/bid-build). What the study demonstrates, more than anything, is that different projects and different owners have different viewpoints of the contract formats available.

It also points up something else: there are a lot of sources in the industry—the *ENR* is one of them—from which an owner can draw useful information with which to make intelligent pro-active decisions, rather than just react to problems or unsupported assertions of benefits.

Design-Build with Owner as CM. There is one major abuse of the design-build concept that's been around for some time, and contractors have learned to be more careful than they first were to get involved in some of these relationships. The owner, in addition to executing design-build contracts with separate subs, acts as its own Construction Manager.

If the owner is knowledgeable, honest and fair, it has been productive. But the owner has sometimes, without adequately defining its Program, continued to pile more demands onto the "design" component of a particular design-build subcontract, until subs in this role have been put out of business.

> For example, the electrical sub, working with its design consultant, who believed that it had a competent scope of work, provided a lump-sum bid accordingly. Only afterwards would it learn that it was also to provide public-address systems, security systems, atmosphere-lighting systems, etc., which the owner, consciously or otherwise, "forgot" to put into the original Program. Under pressure to get paid and in order not to be in default, these subs performed the extra work and then found that the costs of this uncompensated extra work drove them into bankruptcy.

19

These malfeasances aside, Design-Build, under negotiated contracts with known and qualified parties, is proving to have considerable merit, and considerable impact on the way things are done in private and semi-private construction today.

The owner who is undertaking this kind of contract for the first time, however, still needs to be aware enough to ask some challenging questions about Program and budget as part of his negotiations. He also needs to have competent judgment, and legal advice standing by. The need for the owner to invest in a clear Program and Plan is as important for considering D-B as for any other form of contract.

They are still "strangers" out there.

Cost-Plus-Fixed-Fee (C+FF)

On a C+FF contract, the design usually proceeds to some early stage of Design Development. Then, the contractor is retained and works alongside the architect for the balance of design, contributing input that will confirm the buildability and budget of the project.

This type of contract differs from Design-Build in that there is usually a separate manager for the project. The architect and contractor each contract directly with the owner.

The contractor's contract is usually based upon a competitively bid fee (say, 3, 4 or 5% of the total construction cost) limited by a GMP after the contractor has worked closely enough through the Design Development to establish projected cost limits.

The balance of the work (say the remaining 96%) is often competitively bid, with both the contractor and owner contributing to the bid lists for various subcontractors. The construction and procurement are coordinated by the general contractor.

In many ways, this type of contract has substantial advantages for the owner. The owner is able to work closely with the architect in the early stages of design. Later, having established an effective working relationship with the architect, the addition of a carefully integrated contractor brings the insights essential to assure a well- coordinated, buildable, cost-effective project.

Fast-Track

Fast-track construction describes a condition under which the early stages of construction proceed while design continues on the latter and finish phases of the project.

The purpose of fast-track is to gain the benefits of earlier occupancy, reduced carrying costs, reduced overhead and other potential benefits.

It can also be hazardous for the owner.

Fast-track must be associated with a GMP (see below) as a limit for the total cost of the project.

The challenge for the owner is to define the scope and criteria with enough particularity to make the GMP meaningful and reliable.

The problem—often unforeseen until it's too late—is that latter components of the project may not fit, or may not have the adequate design to support them, or early components may have to be reworked to accommodate components that are designed later—all at costs that can easily void the GMP limit.

The major concern of fast-track comes as you're on a downhill grade—the heavily loaded cart (everybody trying to get everything together to meet a deadline) can run over and kill the horse.

How *Program* Relates to *GMP*

It is particularly important if the owner is to proceed with a Guaranteed Maximum Price (GMP) construction contract, to develop a comprehensive Program that incorporates standards of quality and that estimates reasonable quantities of significant materials. The GMP is often developed on the basis of incomplete drawings that do not yet show the final details for such items as equipment, reinforcement, hardware, lighting, distribution services, and so forth.

More than one project has had its budget blown away after the GMP was set, because the contractor discovered that *"the final design incorporated more reinforcing steel"—or more something*—than he *"provided for"* in his GMP.

This is a particular kind of hornswoggle that can be avoided by incorporating definitive descriptions, quantities, units and other reasonable limits in the outline specifications.

287

19

Final Cost—Original GMP Times 150%

The company's empire was so large, and its project so expensive and expansive that it was blinded by its own image. It had signed standard AIA agreements with the architect and with a contractor-CM to design and construct a $110 million fast-track facility under a GMP contract.

A few months into the project, the owner became concerned that the cost of early foundation work was being billed at 1-1/2 to 2 times the costs in the contractor's budget. The contractor assured the owner that the excess costs would "be recovered later in the project." The owner also was concerned that the meeting minutes, maintained by the contractor and architect, appeared to leave out problems that had been discussed but were seldom resolved in the meetings. The owner's remarks and concerns were missing altogether.

The owner discussed the situation with an independent consultant who asked to see the separate contracts, the meeting notes and the progress payments. It took the consultant half a day to determine that the contracts would hold as much water as two nesting colanders. There were no requirements for reporting of design production, of problems, of proposed changes, nor of projected costs.

The consultant advised that unless the owner required some objective evaluation of design progress and actual costs, the project could run to $140 million before it was completed. The owner grudgingly acknowledged that the A/E and CM probably had a greater vested interest in working together well into the future than they did in their relationship with this owner.

Since the contracts were already in place, the consultant could only recommend that the owner introduce him into successive progress meetings. He would only say, "Good morning" and "So long," but after each meeting, he would provide the owner insights about questions that should be raised and issues that should be resolved in successive meetings.

The owner proposed this arrangement in the next progress meeting. Both the A/E and contractor-CM raised such vehement objections to having another party in the meetings that the owner backed off altogether.

The consultant was right in his projection and wrong in his estimate. The project blew right through the $110 million GMP *and* his projected $140 million, to a final cost of $165 million.

The Project Program of a fast-track project becomes the single most important document to keep ahead of the power curve.

Just by way of example. If the work proceeds at such a pace that concrete walls are placed, or sheetrock is installed, before all the equipment and attachment points have been defined, laid out, and templates provided, the owner just might be paying for jack-hammering or ripping out sheetrock in order to mount essential equipment six months later.

For a fast-track project to be successful, someone with both a broad and a detailed view of what must go into the project must ensure that this information is listed, defined, dimensioned and quantified in advance, so it can be used to direct and to confirm the production of the drawings and specifications.

A couple of years ago some genius invented Flash-Track. The owner whose project is flash-tracked will need to ensure there's a competent clairvoyant in charge, to keep his project from self-destructing.

Guaranteed Maximum Price (GMP)

Sometimes also called a Guaranteed Outside Price contract, this is more a description of a limit established at the outset of a project when the final design and/or the hard cost of all the work has not yet been established.

The owner looks to the contractor to exercise its experience at some intermediate point in the development of the design and/or construction to provide a limit to the amount the final project will cost.

This budget-limit concept may be associated with a segregated-prime, a cost-plus, a fast-track or any number of other owner-contractor relationships.

It is important, as with any but the lump-sum set of construction documents (which are presumed to completely reflect the owner's requirements), to define in some supplementary way what has not yet been designed or specified. Sometimes, it may require the owner to develop a more definitive program. Sometimes, to spell out certain performance criteria, or quality criteria, or even to establish quantities for certain materials.

The important issue is that a GMP can only be meaningful when what is not yet fully detailed in the construction documents is otherwise defined.

19

Design-Build: Yes, No!, Maybe?

A number of studies have recently compared Design-Build with other construction-contract formats. It was found that there were only minimal overall differences in costs, claims, durations and quality.

The District of Columbia, having found that the initial costs of Desing-Build were too high or non-responsive on its new convention center, now seeks an At-Risk CM.

Meanwhile, the City of San Antonio has bought off a CM that, without responsibility for final costs, had wangled a front-end bonus contract on the "savings" on bids for a $110 million convention center. The City now asserts that it will look for *independent, objective assistance* before developing requests and contracts for future services.

And the Federal Bureau of Prisons (FBOP) is shifting to Design-Build because it believes that, while there are some additional risks involved, and the initial costs will be higher than lump sum jobs, D-B will reduce change order and claims problems. The agency's design and construction chief (hopes) *"that [the] premium [the agency pays] is a manageable amount."*

The FBOP has retained *three firms that specialize in D-B services* to prepare the documents for its proposal requests.

Who knows what lurks in the hearts of folks when phrases like *"independent and objective," "manageable amount"* and *"some additional risks"* are the standards used to retain those who will first define their contract work scopes *and then are likely to bid on them?*

Is it the same that's found in the AIA A201: *"the Architect will **endeavor** to secure faithful performance... [and] will not show partiality..."?*

8

NEW TRACKS

Borrowing some values out of the past and evaluating your own capabilities and commitments to implement an effective Plan.

20

TURNAROUND

This is what we know:

- Successful building projects are a direct reflection of the respect an owner generates with those who design and build them.

- Respect flows when an owner has clearly defined his goals, has incorporated them into construction documents that are buildable, and administers those documents with honesty and fairness.

- To the extent those conditions are consistently achieved, an owner can avoid 80% or more of the problems that beset many projects and ruin budgets. He can reduce the number of "other strangers" standing by to invade the territory of his project.

- To accomplish these goals the owner needs to be able to measure the design performances of those who produce his construction documents.

- And during construction, the owner needs to be able to measure and document the interactivities between those who produced the documents and those who are building from them.

When the Owner Was in Charge

For most of the 4,000 years of recorded history—right up to the middle of the 19th century—the owner was in charge of his own building projects. Whether through Draconian measures, hereditary right, or his own capacities, the owner controlled the function, design, resources and construction of his own project.

The Greeks left us examples that are models for us today. The Romans built not only monuments, but the first practical highways, aqueducts and successful commercial buildings. As Europe left the Dark Ages behind, the master builders both designed and built, at their owners' direction, marvels of light and beauty that still inspire us.

20

The Romans sometimes directed their architect-engineers to stand under their structures as the scaffolding was removed.

The hereditary leaders of the Renaissance directly controlled the construction of buildings whose design and workmanship continue to call us from halfway around the world to learn from them.

The owner was in charge right up to about the middle of the 19th century. By then, however, new forces were already in motion to change the owner's role and to intimidate him into subordinating his ideas to the control of others.

The Industrial Revolution was accelerating. It allowed us to span greater spans, reach taller heights, incorporate new materials and equipment into our buildings. Construction became more complex. It required the integration of new disciplines in both design and construction.

In 1888, the AIA, then only 33 years of age, joined with the American Builders Association to divide up this increasingly complex territory. The contractor would manage the construction and his subcontractors. The architect would manage the other design disciplines *and* the contractor. The standard AIA contracts, with the architect as project manager, were born.

The architect's claim to management was staked on two premises: he was there first, and he could produce pictures of the final project. (Drawings of electrical circuitry or mechanical systems are not persuasive selling tools.) Besides, he was the descendant of the *master builder* and the *arketecktron* of the Greeks.

The architect would integrate the multiple disciplines. He would prepare the budget. He would define the construction duration. He would deal with the contractor. He would resolve disputes between the owner and contractor.

The owner became progressively more separated from the control of his project. Once he outlined his needs, the owner was nudged out of his role as *leader*. The architect would lead and manage. The owner was required only to pay the bills.

This realignment of the management relationship was the result of one of the more ironic presumptions of the construction industry. That the architect's attributes—being there first, and the ability to draw—were measures of *management ability* and of *leadership qualities*.

This realignment of the management relationship was the result of one of the more ironic presumptions of the construction industry. That the architect's attributes—being there first, and the ability to draw— were measures of management ability and of leadership qualities.

Advances in industry, two worldwide wars, extraordinary leaps in technology and the demands of a socially and environmentally demanding culture have reinforced the owner's sense of separation and intimidation. Few owners today comprehend the resources required to build their projects. It's a separation that's been reinforced by the numbers of strangers who now come into the owner's project territory.

As if more separation and intimidation were needed, the law's discovery of construction as a fertile soil for expansion has driven the owner to seek shelter with the ostrich. Most owners carry into their projects the often-misguided hope that those who understand the law, and those he hires to design and build his project will be looking out for *his* interests.

The 1,000% increase in construction claims from 1965 to 1995, that we've noted before, is evidence of how the distance between the owner and his project has grown.

This increase in claims has been paralleled by the exponential growth both of the panaceas to cure their causes and the cottage industry that reaches out to exploit their "resolution."

20

Over 150 years of this growing separation from his project, the owner has been led to believe that he was being protected by the same profession that originally encouraged the separation and proposed to manage the pieces. The owner has been told—*contractually*—that the architect would exercise his fair and impartial judgment to settle disputes.

And when—inevitably—the architect's fairness and impartiality came into question, the owner has been led to believe that only *the law* can be the arbiter of how agreements are written, read and interpreted.

The owner has been led to believe that he is required to deliver his Program and then stand by passively, with adequate funds, but without appropriate or adequate means to measure how those funds are spent. He has come to believe that it is beyond his need or competence to define, measure, document or influence the performances of those he hires to design and build his project.

Taking Slices from History

There are two slices of history—together with the evidence of successful projects you've found scattered through this book—that should move an owner to shake off this passive role. They demonstrate that an owner only needs to recognize his own capacities for leadership to reassert control over his own project.

History reports that owners who exercised Draconian measures—dismemberment, banishment or burning at the stake—were able to control their projects, and pretty much avoided *all* of these problems, and *all* the unwanted strangers as well.

> *But our culture frowns on this approach, so those measures are out.*

History also tells us that when the master builders got too big for their britches, the popes excommunicated them.

> *But there's no longer much clout in that approach.*

296

During the Renaissance, the Medicis and their equals exercised both aesthetic and fiscal control over the entire life of their projects.

But that required them to have training in the arts, a purse and a leisure we don't have today.[1]

Besides which, the Industrial Revolution and modern technology have made it difficult, not to say impossible, for the owner even to try to comprehend the multiple disciplines required to design a building in today's environment.

But there are two periods that provide productive images. One of them comes out of the 17th and 18th centuries and culminated in the Declaration of Independence.

The other period goes back farther—all the way to prehistory.

In the 17th century, man woke to the possibility of questioning his surroundings—a possibility from which he had been kept by the intimidation that had surrounded him for 1,500 years.

The discovery that there were two continents and a second ocean between Europe and the Far East precipitated a transformation. It led within a hundred years to the Age of Reason.

The central themes of that age were *questioning, measurement and objectivity* as standards for reasonable thought. They displaced abject passivity.

In terms of man's future, these standards were lucidly expressed by Tom Paine in **The Rights of Man,** in which he concluded:

"…it becomes evident… that a general revolution in the principle and construction of governments is necessary."

Substitute for "governments" the words "building projects" and Tom Paine speaks for the *Rights of the Perceptive Owner* today.

But how can an owner initiate a revolution in the design and construction processes of his building project if he doesn't even know how to use the tools of measurement?

---

[1] *Except in rare cases such as the recently completed $1 billion Getty Museum.*

How can the owner of a construction project become *the leader* when he can hardly read the squiggles in CPM schedules, understand the lines in the documents he has bought, keep

up with current regulations, or interpret the words in the contracts he signs?

Farther back in time—about 20,000 years ago—the tribe followed its *leader* out of the caves into the fields, built communities, planted crops, developed domesticated animals and generally began the civilization we know.

In addition to the leader's energy and strength, these activities required a direction—a sense of rhythm and order. It was the leader's synergistic relationship with the *shaman* that made this order possible. It was the shaman who interpreted the stars and the moon, traced the movement of game, the winds and the seasons and read from the magic pictures. He advised the leader where to hunt, when to plant and harvest.

The Owner's Options—Your Options

Build It Twice challenges you, the owner, to shake off the intimidations that have beleaguered construction in the last 30 years. To help you do that, *Build It Twice* has reconstituted the shaman as *the owner's guide.*

Shaman—guide—someone who has been down the road before. Someone you trust. Someone who knows how to read the signs and is prepared to help you—the *leader*—to interpret and to follow the signs that will deliver a successful project.

Your immediate need is to develop a Plan—a scheme for harvesting and administering a set of buildable construction documents. To do that, you need to develop familiarity with the tools that will be used to implement that Plan. For that you will need to develop the same synergistic relationship the *leader* had with his *shaman*.

Working together, your purpose, first, will be to develop familiarity with how to read and work with the tools that will be used on your project. Then, to develop a Plan through which you will *delegate, measure, question and document* what others will do with those tools to develop your project.

As *leader*, it won't be your role to interpret the codes and regulations. Or to manage all those disciplines who will incorporate them, together with your Program, into your documents. Or to manage how they're implemented during construction.

Your role, with your guide's support and assistance, will be to shrug off the friendly intimidations of architects and attorneys, to tell *them* what scopes of work you intend to delegate to them, and to ask the knowledgeable questions that will confirm whether they can perform to standards you and your guide have established in your Plan.

But owner—even before you go looking for your guide, you need to confirm that you're prepared to be the leader, one that both your guide and those strangers you hire will respect and follow.

21

THE OWNER AS LEADER

For all we've said about *project success,* the prospect of that success comes down to whether the owner can demonstrate certain qualities to those he hires—qualities that will motivate their respect.

Forty plus years of working with, around, between, for and against them has demonstrated that there are four broad categories of owners. Each generates its unique consequences on a building project.

Owner Profile	Consequences
Knowledgeable and involved	Produces the best results for both the owner and the project—by far.
Knowledgeable but detached	Not as good—can create misunderstandings and delays in decision-making—but better than what follows.
Naïve but detached	Dangerous for all concerned. A recipe for loss of control over the prosecution of the project.
Naïve and involved	A recipe for disaster.

These four categories only hint at how the disposition, awareness and involvement of the owner can influence the success of a building project.

When you color these owner qualities with timidity, bureaucracy, fear, politics, ego, diffidence, opportunism, self-aggrandizement or favoritism, the prospects for respect or success for even the first category of owner become dismal. Add them to the other three categories and catastrophe is almost guaranteed.

21

So it becomes important for an owner to consider what kind of glue it takes to seal out these destructive elements. For the owner to be a *leader,* not just a follower or a martinet.

Glue

There's no getting around the fact that construction is a potentially confrontational environment. The owner must be prepared to meet confrontation with reasoned energy.

When energies are dissipated by resistance from within the owner's own organization, even the most determined leader is likely to need some fortification.

The kind of fortification—glue—that's to be found in **William Bennett's** *Book of Virtues:*

> *"To be honest is to be real, genuine, authentic, and bona fide.... Honesty imbues lives with openness, reliability, and candor.... Honesty is of pervasive human importance... every human enterprise requiring people to act in concert... is impeded when people aren't honest with one another."*

> *"'We become brave by doing brave acts' (A quote from Aristotle)... Threatening things is not to be confused with fearlessness. Being afraid is a perfectly appropriate emotion when confronted with fearful things. The infectious nature... of courageous behavior on the part of one person can inspire... a whole group."*

In case you believe the ideas expressed by **William Bennett** are fantasy in today's world, try on the words of a handful of business leaders out of several hundred who have spoken about the relationship of respect to honesty, awareness, leadership, fairness and integrity.

Raymond W. Smith, Chairman and CEO of Bell Atlantic:
"One of the most widely reported problems in American business is the lack of clear, corporate direction... as a guide for action."

Donna Shalala, Secretary of Health and Human Services:
"The core values [of] a successful leader are honesty, sensitivity to other people's needs, and a... commitment to fairness... it takes courage to be a leader...."

Robert L. Crandall, Chairman and President, American Airlines: *"You have to behave in a way that people can be proud of. You've got to have the highest ethical standards to be an effective leader."*

Jack Welch, Chairman and CEO of General Electric: *"...one concept we have in [GE's] culture is the idea of setting the bar far beyond what you think is realistic... the results often will blow your mind. Coaching is another key to [development]."* *(Shades of the guide we'll meet in chapter 22.)*

The Owner as Leader: Others Will Manage

The first consideration for the owner is to distinguish his role as leader for project development from that of the various managers who will be hands-on for administering design and construction.

The role of the *leader* is to set goals, define limits and establish criteria for performing the scopes of work required to deliver the project.

The role of a *manager* is to interpret those goals into scopes of work and to set the forces into motion to get the work done, to be proactive in measuring and anticipating events competently, and to report to the leader the implications of what he sees, what's being done and what actions he recommends be taken.

The leader will accord a competent manager with decision-making authority consistent with his abilities and the confidence the leader has in his integrity and objectivity.

The leader will know where to draw the line for the ultimate decision-making authority he will retain.

The suggestion that the owner needs to become familiar with the documents, tools and reports of contract management is not designed to prepare him to be in the trenches on a continuing basis.

This self-education is intended to assist the owner to develop the instincts to know when he's being "opportunized," To enable him/her to see things coming when others may be glossing over them, to equip him to understand sufficiently the difficulties of construction so he'll be able to communicate with and to motivate the strangers he retains to visualize the project goals as clearly as he does.

21

There's Homework to Be Done

You will need to set up your own study course to cover the blank spaces in your own personal *gap*.

If you're an experienced owner, this may require only refinement of the criteria with which the next project team will be selected, modification of the requests for qualifications and contract requirements, and establishment of procedures to ensure adequate measurement and documentation.

If you haven't been down the project development road before, or if you've been disillusioned with past performances on other projects, you will probably want to set aside several weeks, or even a few months to ensure that you have a Plan that will work for you.

Here's a quick recap of the resources you should be equipped with in order to define an effective Plan:

- Sufficient familiarity with the standard design and construction agreements (starting with the AIA documents, say, as a baseline) to modify the requirements in these agreements and specifications with the assistance of your guide and an attorney who shares your perspectives.

- A clearly defined system of procedures and systems with which to measure the performances in those agreements.

- Sufficient familiarization with the content and organization of construction documents, including drawings and technical specifications, to be able to read and challenge the progressive development of the documents on your project.

- Sufficient familiarity with budget, cost and scheduling information, including network analyses, and with the level of detailed information required for budgets and schedules to be competent bases for measuring performances.

- General familiarity with the processes of design and construction, and with the interactivities between the design team and the contractors, and the paper flow these interactivities will generate during construction.

Recapping Earlier Chapters

We defined project *success* two ways in chapter 1.

- One definition of success was related specifically to the owner's goals.

- The other definition related the optimization of all of the project's resources to avoidance of the black hole of a construction claim.

In the intervening chapters:

- We visited the strangers with whom the owner will need to work to produce his or her project.

- And met some of the "other strangers" the owner would just as soon not meet as camp followers.

- We referred to a tool—the Buildability/Budget Analysis—with which to improve the production of construction documents.

- We suggested the need for the owner to span the information *gap*.

- And we indicated that all of these would come together in the owner's Project Plan.

It's time to start organizing *your* Plan.

How to Identify a Potentially Successful Owner from an Unsuccessful One

Even before the Project Starts:

The successful owner develops a project-leadership perspective, knows that design and construction are not democratically directed enterprises, develops a mind shift away from daily activities and sets aside time for planning, decision-making and leading that is distinct from the time for his/her "normal" activities.

The unsuccessful owner does not set his/her "normal" activities aside sufficiently to recognize the unique demands of an unforgiving, time-driven enterprise, is bound to the democratized, committee-driven mores of decision-making and does not understand the essential relationship between "delegation" and "measurement of performance."

The successful owner itemizes his/her own capabilities with regard to the unique environments of design and construction, determines where he needs assistance, and defines a scope of work and parameters to consider how he might find that assistance (the guide and attorney we've been discussing).

The unsuccessful owner, recognizing some limitations of perspectives or time, finds a crony to "help manage" the project development.

The successful owner sets aside the time necessary to develop a competently defined *Program* and *Plan,* and projects these ideas through a series of *what-if exercises* sufficient to prepare himself/herself to make timely decisions as the actual design and construction advance.

Even before the Project Starts (continued):

The successful owner delegates definitive assignments to his guide and his attorney for them to prepare the procedures, agreements and outline specifications required to put his Program and Plan into place with effective tools for measurement and documentation, and uses the perspectives developed through this effort to manage, direct, lead and motivate others in his organization or agency to "get aboard" a program that has a realistic potential of delivering a successful project.

The unsuccessful owner goes directly to strangers for advice, without competent parameters for their selection; allows them, or the established legal hierarchy of his organization or agency, to dictate the terms of their retention; and delegates assignments without adequate scopes of work, and with no sense either of the processes that will ensue, or how to measure them.

The successful owner knows what kind of questions to ask to avoid problems. When problems do occur during design or construction, he/she is prepared to deal with them quickly; is able to negotiate effectively; and has a reasonable expectation of completing the project without experiencing a construction claim.

Meanwhile the unsuccessful owner, when the almost inevitable problems come up during construction, is always *reacting*, finds his only recourse is *denial*, and is totally unprepared when a construction claim cripples his project.

9

YOUR PROJECT PLAN

The resources you will need, and some approaches you might consider, to produce a Project Plan that will deliver a successful project.

22

YOUR GUIDE

Even when you're doing something familiar—on the golf course, in the kitchen, in the gym, enjoying music, or planning a trip—having somebody there to compare notes with can contribute new values to the experience.

When the trip is to a new and strange country, you know you can avoid a lot of delay and awkwardness by discussing it with someone who's been there. The longer the trip, the stranger the country, the greater the need for preparation.

You wouldn't consider climbing the Matterhorn without a guide. For planning a trip like that, three months isn't too long. Six months, or even a year, would be better. Several months of easy, evolving conversations with someone who's been there would be ideal. You'd be able to roll around the ideas you've collected in your travel notebook.

If you think about your next project as an adventure and you can find a guide who contributes his or her own enthusiasm, it's surprising how many new perspectives you're likely to discover. You'll have someone with whom to discuss the questions you've collected in your project notebook.

Since you'll be making changes from the conventional ways in which most projects are approached, you'll need as much time as you can find to make adjustments in your psyche, or your organization.

If there's a project on your horizon, it can't be too early to find your guide, to start the homework and to find the resources you'll need before you invest in your project.

One of those resources is the construction-knowledgeable attorney whose attributes we've already discussed in chapter 11. You'll need to know more about the country you're travelling to before you talk with him anyway, so this is the time to consider the profile of the guide you'll want to find.

22

Profile

There will be plenty of people who will tell you that the ideas in this book won't fly. If you spend too much time with them you might as well throw your notebook away.

What you're looking for is a pragmatic idealist. A *can-do* person who's also a realist. If you've come this far, it's not a matter of *if,* but of *how.*

■ He or she has been part of, or worked closely with, the resources that produce building projects, has been through the problems those processes can produce, and close enough to those problems that he doesn't find many surprises in the early chapters of this book.

■ You both find there is a certain positive chemistry that will produce candid, productive judgments when you discuss how you intend to administer contracts. You need someone who sees possibilities where others see only problems, and will be productively honest and candid with you about all performances, including your own. You need to be entirely comfortable exposing your ignorance with him/her.

■ He or she needs to have the capacity to translate pictures and words (drawings, specifications, contracts) into images, and into practical ways to get things done.

■ He/she needs to be able to *conceptualize*—to project potential conditions and relationships into the future. This capability is crucial if you're going to anticipate problems and to respond to them before they get out of hand.

■ He needs to be sufficiently self-reliant and mature to have neither an axe to grind nor an unsatisfied ego— because you may continue to need his help later after you put your Plan into action. He/she has already met his own goals and is willing to share his experiences with you. He's not on an ego trip of his own, but has the fortitude to say what needs to be said with candor.

■ Especially if you will continue to rely on his/her judgment as your project evolves, your guide should have the people skills to spend at least 50% of his time listening, 25% more of it thinking ahead of things, and usually moves his mouth only when it might be productive.

■ He or she should be the kind of person you'd want as a fishing guide, a golfing coach, a good camping companion or a guide on the Matterhorn.

- Your guide should be a left-brainer. Planning, calculating and measuring are left-brain activities.

- Your guide must be totally independent of any contractual performances that go into producing your project.

The guide you're looking for has some of the characteristics of a family doctor—a doctor who thinks preventative medicine in terms of the health of your project.

Experience has confirmed that in some organizational environments it is useful to maintain a confidential relationship with your guide, at least until your Plan is adopted and set into motion. This is not out of need for secrecy in the long run, but so that the distractions and roadblocks of the naysayers will be put in abeyance until you're satisfied that you've worked out a Plan that will fly with your organization (More on this subject in chapter 25.)

Your relationship with your guide may continue only through the development of your Plan and your Program. He or she will help you to set up the procedures you'll need to measure contract performances, work with you and your attorney to put your ideas in agreements and specifications, and may then go to standby.

Once you have a workable plan in hand, you may decide to go it alone, delegating the administration of those contracts to others—a CM, a design/build contractor, or some other type of project representative.

You may want to keep your guide in a standby mode, available to assist when new conditions come up for which you would benefit from his or her experience.

Or you may request that he or she be available to assist with the periodic B/BA reviews of the developing design documents.

Or you may even resolve some kind of retainer agreement, so that your guide is available at pre-scheduled times to assist with evaluating the reports you receive, and to help organize your document archives and database.

22

Compensation

You should not expect your guide to be free. Nothing free maintains its value for long. It's in our nature that the worth of the services you receive requires some demonstration of satisfaction. You just might need your guide's assistance down the road. The manner in which you establish his or her standby mode will establish his or her responsiveness if it's needed later on.

It's impossible to discuss rates, though, out of the context of particular environments. What may do in Sioux City or Santa Fe will be entirely different from what it takes in Seattle or Baltimore. Added to which, the more complex the project, the more experienced your guide will need to be. The time he or she will be able to devote to your concerns will also be a consideration.

Your guide may be a firm instead of an individual. But it should be a firm in which you will have the assurance of the principals that you can rely on their confidentiality—that, and a guarantee that there will be no conflict of interest with those you hire later. You may need your guide's objective judgments when problems arise.

A Building Project Owner's Association

The amount of homework you may need to do and the matter of compensation suggest another potential.

There are certainly others in your community or industry who have projects on the horizon.

As long as you are able to separate your own decision-making when the time comes, from the negative responses of others, you might find it in your mutual interests to organize a local Owner's Association for the specific purpose of learning about the tools required for effective measurement of performances: cash flow analyses, schedules, budgets, payment reports, project documentation.

This kind of informal association could structure a curriculum, and would spread the costs of the learning process you need to undertake. If you're able to encourage a group of people with similar goals and levels of experience, it could facilitate your familiarity with the procedures you'll put into your Plan.

It will also provide the opportunity for experimenting with *dry runs* of the ideas you intend to put into your Plan.

Such an association might even, over time, demonstrate to your local design and construction firms that there is some serious stuff going on among owners in their community. If that occurred, you might find there is some resistance out there, but by and large, the construction side of the industry will welcome more knowledgeable owners (group 1 on page 301).

One caveat, however. Associations like this can contribute new insights, but they can also contribute negative vibes. You will still have to develop *your* Plan for *your* Project once you're comfortable with the tools of management.

The Value of Knowledgeable Questioning

You will find that if your guide has the attributes above, the most valuable contribution he or she will make will be to ask questions that will help you refine your ideas and to test them against reality. Plan to develop *what-if* situations that both of you will work through to practical solutions:

> *How can we ensure that the design consultants will understand what our Plan requires? Are the requirements of our Program clear and comprehensive enough to be able to measure design as it evolves? Do we need assistance with the cost estimating?*

> *Who can we consult with to start the development of our preliminary Program, so we'll be ready for the architects when we bring them in? What do we put into our RFQ for design services? Let's define our plan for backchecking on the design consultants' track records.*

> *How do we incorporate the procedures we've defined into contract terms? Do we need an outline to explain them to our attorney?*

> *What resources will we need to follow through with our progressive B/BA milestone checks?*

> *Let's discuss integrating my Plan into my organization. Should we test it with a dry run so we can anticipate some of the internal objections I'm likely to hear?*

> *Let's project the typical responses we can expect from contractors. What's the best way to ensure responsive bids in this current environment? What responses are we likely to get from sureties?*

22

What contracts and specifications should we use as outlines to develop the design and construction agreements?

A Discussion You Might Have With Contractors

In view of the firms out there that will challenge your right to do things differently, it will take some fortitude for you to carry it off.

Once you have authorized your architect to start work, you might consider making contacts with the construction industry. Filtered through the experience of a perceptive guide, the industry can make constructive contributions to your project, even if you intend to issue it for competitive, lump-sum bids.

The construction community, when consulted in advance of bidding a project that raises their interest, can be surprisingly willing to contribute constructive suggestions.

On the next page is a bill of materials you should ask your guide to collect for you to use with the curriculum in the next chapter.

Bill of Materials
Equipment

- A roll of scrim paper, 18" or 24" wide.

- An architect's scale (an engineer's scale is optional).

- A straight edge.

- A Dundee marmalade jar of No. 2 or "F" pencils. (There's something satisfying about having a full jar of sharp pencils at the ready. An electric pencil sharpener is also good.)

- A large Pink Pearl eraser. (If you can borrow an electric eraser from someone, you'll be able to change your mind even quicker.)

- Access to a computer with, at the minimum, scheduling and spread-sheet software. A word-processing program and a flat-file database system would also be useful as you refine scopes of work and specifications. (The size and speed of today's personal computers are a boon. In 1960, computers and printers were klunky and filled a room the size of your kitchen.)

* * *

These are the materials you may want your guide to help you assemble:

- An AIA B141 and AIA A201 standard agreement. The 1987 is OK. Though it's been updated, many projects are still being designed under the '87 version. From both an owner's and contractor's viewpoint, the later version only exacerbates the shortcomings of the '87 version. And you're only using it as an outline anyway.

- A set of construction drawings and specifications from a project similar in size to the one you plan to build, preferably for a building that your guide can walk through with you as you do your homework in the drawings.

- A collection of daily reports, inspection reports, transmittal logs, a typical shop-drawing submittal, RFIs, a change proposal, change orders and field-proceed orders. Payment request forms and meeting notes.

- Cost reports, budgets and quantity take-offs.

- Several examples of bar charts, 3-week-look-ahead schedules, CPM schedule network diagrams (real ones) and schedule reports.

A Project Journal—Specifications

One of the most valuable pieces of documentation an owner can have maintained on his or her behalf is a scrupulously, objectively maintained journal of activities, events and conversations on the project.

To have value, this record needs to adhere carefully to these guidelines:

1. It should be as regular as possible. The closer it is to a daily record, the more valuable it is. (See the end of the story on page 249.)

2. It must be maintained in a bound book, to establish authenticity and to underwrite its admissibility (see the glossary). As an alternative to this, copies of records maintained on loose leaf or computer database records can be authenticated by sending them regularly to the offices of someone whose independence and integrity will remain unquestioned and who will carefully date-stamp each page of the records as they're stored.

3. They should be as complete as possible. To consider what this means, see pages 53 - 55, and page 197 on the content of the daily reports you will require the contractor to maintain. Your journal (maintained by that objective, independent person you rely upon to keep you informed in the course of design and construction) should record conversations in meetings, events on the project, equipment and body counts (by trade), or any other facts that could be relevant to the performance of any parties you have under contract.

4. *What your journal should not do* is express subjective reactions. *("The contractor's doing a lousy job....." "The architect doesn't know what he's doing.....")* These comments will carry no weight—in fact, they'll have a negative effect—with a trier of facts. It needs to be as complete, detailed and objective a statement of what's going on at any time as your record keeper can make it *without subjective judgments.*

5. *No barracks-room attorneying.* Think of your record being read by the man in the white wig behind the high bench in the Queen's Court. He will have no patience with any sort of legal interpretations. *Only the facts, Ma'am.*

One caveat. If you, the owner, aren't performing, either forget the journal, or be sure to record your own failures for it to have credibility.

23

THE TOOLS

You may be approaching the prospect of a *study program,* or *re-entering the learning curve* with some reluctance. We remember our first time on skis, our first dance, our first speech, the first time up on a bike. What lingers may be the embarrassment, or the fear that we can't learn new things.

But the other side of the memory is how quickly our ignorance and awkwardness disappeared with Dad's help, the assistance of a friend, an older brother, a coach or a guide. That's one of the reasons it's important for the guide to be there: to reduce the length of the learning curve.

There's one major difference between your first golf swing or bike ride and your project: you'll only get one chance on your construction project. The costs of learning on the job how to use the tools you'll need are prohibitive.

To accelerate your learning curve with these tools, consider applying them first to something you already know well. Something you might share with an associate or you family, like scheduling and budgeting a vacation or a large party.

The tools we're going to concentrate on are the ones you'll need to conduct the Buildability/Budget Analyses during the design phases of your project.

Since the purpose of the B/BA is to anticipate the views of the contractor, these same tools will be basic to the documentation you will require others to produce during construction.

The Buildability/Budget Analysis

Once we cover this part of your Plan here and in the next chapter, and you know what it's intended to accomplish, you can call it a Constructability Analysis or anything else you want to call it. You will know the kind of detail that will allow you to write a scope of work to obtain the results you will want from the design process.

The B/BA works backwards from construction, like the organization of this book. That's because it's intended to reflect, while your project's being designed, the judgments contractors will make.

23

Set Yourself Initially to Have Some Fun with These Tools

We've already said some pretty serious things, and there will be some serious considerations coming up once you're comfortable with the tools. Even if you're looking at a near-term project, though, your facility with them will come much faster if you apply these tools first to something you already know—like planning a vacation, a birthday party or a workshop project. All of these projects will offer good applications for both scheduling (CPM) and for accumulating costs (computer spreadsheet).

Ask your family or friends to participate. Distribute project assignments around the group (subcontractors and suppliers). Make it a game. Awards go to the best suggestions for shortening the time. Include buying materials, organizing outside deliveries. Divide up responsibilities.

After the first time through, you'll quickly find things you overlooked. You'll start breaking some activities down into parts and separate assignments. You'll find you can split up things you originally did end-to-end.

More than anything else, the challenge to improve the overall duration you start with will demonstrate that a CPM network isn't hard to work with. You'll also find that an eraser is much more satisfying than the delete key on a computer. Ideas flow out of a pencil that don't connect with your fingers on a keyboard. It's also satisfying to see your ideas in your handwriting on a piece of scrim paper.

When you get your guide involved to put your data into a computer scheduling program, you'll see that all the work has been done. The computer is just a dumb adding and printing machine that can be operated by anybody who's taken a 3-day software course. You have found one of the first places to delegate a chore on your project.

When you experiment with a database or spreadsheet, use it first to inventory some things around the house, or the stuff in your garage or your office. Associate numbers with the things you find there and you might even be surprised how much you've spent in The Sharper Image or at Grand Auto over the past few years.

When you sit down to learn something about reading construction drawings, keep your scrim paper and eraser nearby. You might want to try sketching what you're trying to visualize (conceptualize?). Later you can drive by the building the drawings were used to build and see how good you are at translating flat drawings into three dimensions. You might be better than some architects and superintendents.

Set Yourself Initially to Have Some Fun (continued)

You might spend a couple hours at your nearest technical book store, going through engineering, CPM scheduling and construction-management texts. Just don't get too deep into what you find there; don't buy anything right away. Engineers tend to put everything into formulas, Greek symbols and tables. Those texts are written for the people you will hire to manage your project for you. There'll be time to learn more about what some of these books say later on.

Your concern is not to learn how those folks do what they do, but to develop some familiarity with what they're supposed to do.

But there is one book you should look for: *How Buildings Work,* by Edward Allen, architect and teacher at M.I.T. and Yale. (Published by Oxford University Press, 2nd Edition, 1995.) It's about $40, full of insightful information and useful diagrams.

Through your agreement with the architect, you will receive a definitive production schedule for your construction documents. The schedule will have as its foundation specific descriptions of the products the architect will produce at the end of each design phase.

Then, depending both upon the length of the design process and the complexity of your project, you and your guide will have defined intermediate *milestones* in each design phase at which points the design team will have reached competently defined intermediate stages of those design phases.

At each milestone, you will receive a checkset of the design documents as they have advanced to that point.

From each milestone set of documents you will start to *build your project on paper.* Your first B/BA will probably produce a bar chart and a simple projection of costs.

At each milestone, the detail in your schedule and budget will be expanded from the information in the design documents until, by the completion of the construction documents, you will have a detailed CPM schedule, a cost estimate and a cashflow curve derived from having allocated costs to the work activities in the CPM. Your budget will have been developed independently of the architect's *probable construction cost.*

To accomplish this with your B/BA, you'll need to become comfortable with the following tools.

Bar Charts

We first visited the bar chart in chapter 14. There's not much to add to what you'll find there. It's a good first-stage tool when you don't have much detail to work with. It's flexible, in that you can add tasks, or work activities as they occur to you, up to the point that they become so extensive that it's time to move the information into a CPM.

You will continue to see bar charts throughout the development of your project. Contractors find what they call the *three-week-look ahead* practical a tool when working with subcontractors. Look-ahead schedules on 8-1/2 x 11 paper will be distributed by the contractor at his progress meetings, whether or not he is maintaining a comprehensive CPM schedule.

They're an excellent tool for the contractor. But they're no substitute for the schedule you require to track the project. It is not in your interest to accept them as a substitute. Three-week look-ahead schedules are not adequate for projecting your project sufficiently into the future, for controlling deliveries, for interfacing different disciplines, for measuring payments, or for assessing the status of work that may have been delayed or otherwise disrupted.

It is essential that, whatever tools the contractor uses with his own people, you be provided with a progressively maintained, detailed schedule that makes all those measurements possible.

CPM

You will find an extended description, along with a brief history, of CPM in Appendix A-2.

I've already suggested that you become comfortable with this tool by scheduling something familiar to you. Something you may have done before, without the specificity of listing, sequencing and calculating just how long it took and how many resources you put into it. Planning a vacation can be an ideal project that can involve the input of others (you can use minutes, days or weeks).

You're going to find in A-2 that I've introduced you to the *time-on-a-line* system of drawing a CPM analysis. I encourage you to stick with this method until you are thoroughly comfortable with the logical processes a CPM requires.

The *time-on-a-line,* or *arrow-diagram* CPM is more rigorous than the *time in-a-box* "precedence" system. Carefully drawn, so that lines don't run backwards, it requires you to think more rigorously about relationships. Most of the world has gone to the so-called *precedence* system—and you will be able to whenever you want to switch—it's only a matter of how the computer-process is numbered for network analysis.

But for now, stay with the arrow diagram, complete with *dummy restraints* (dotted lines) and *lead times* (LTs).

You'll just have to take it on faith for now—you'll become much more proficient in projecting your own time, and measuring and testing the time usage of others, if you first understand the *arrow diagram technique* for CPM.

Cost Allocation

As you develop a detailed CPM diagram for your project (and possibly even for your vacation), you should think about selecting and defining your *work activities* in terms of two standards: [1] an activity defined at either end by the point at which one trade (assignment) hands off the work to another trade, and [2] in terms of *cost*.

> The point at which the metal-stud subcontractor turns over a section of the partitions to the electrician for *electrical rough-in* is one example.

By also thinking about cost, you will make it more practical to cost-load the CPM schedule after its duration is calculated.

> As an example, the erection of a steel frame for a building represents only one trade—the ironworkers. But the frame may take 4 months to erect. You need to have an intermediate point for payment. It may be practical to divide the total erection cost by 4, so you have some basis for monthly payment. You may even refine this by weighting the first payment to cover mobilization.

> When it comes to utilities and plumbing, and the overall duration is projected to be 9 months, again it becomes a matter of finding a practical way to measure what will be paid for at the end of each month. It's very difficult to have any assurance about what you're paying for if each month you need to estimate progress against a single item for *plumbing—$453,000*. By resolving measurable allocations on some reasonable basis with a contractor at the

start of a project, you can reduce a lot of time, discussion and disagreement later in the job.

Since you will be working with the B/BA during design, it will be an ideal time to become comfortable with the CPM system without being under the same kind of pressure to make decisions that you can be during construction.

You will also have a product in hand before you go to bid with which to negotiate those reasonable allocations with the contractor after you award the construction.

This is a useful place to discuss front-loading. Under Estimates and Computers in chapter 17 we discussed the potentials for front-loading. You may believe, or have heard, that you can eliminate it with a cost-loaded CPM. Not likely. For the same reasons you read about under Estimates in chapter 17.

Contractors will do what they can to work on your money instead of theirs. And you won't be able to eliminate that proclivity altogether. But you can bring it more in line with reasonableness by having as a resource a realistically cost-allocated cashflow before you go to bid.

Spreadsheet

A computerized spreadsheet system is just about the most universally powerful tool at your disposal. I hope you're already conversant and comfortable with a spreadsheet system. It can help you with your personal budgets, your checkbook, your kids' allowances, your taxes—ad infinitum.

It's the place to start with your project budget.

And just as with the bar chart, you can start simply. Your first spreadsheet allocation may be a one-liner: $368,000. Progressively you can break it down, just as you will the bar chart, as you learn more about the elements of your project.

Gradually, you can add columns, just as you added lines and new categories of work to the bar chart and CPM. The columns can be headed "labor," "materials," "equipment," "overhead," "insurance," "profit"—as many categories as you identify where money will need to go into your project.

The process itself will contribute new insights. You will find "fees," "permits" "environmental programs" and all sorts of items you never thought about as the first images of your project danced in your head.

As painful as some of these additions may be, it's better to learn about them earlier than later.

With the help of your guide (he/she should have at least a modicum of cost-estimating experience), you might even break costs down into "general contractor," "subs," "suppliers" and "consultants," "test-borings," "special inspections," "financing costs," and "code-compliance submittals."

Each definition will eliminate a surprise. Each one will contribute to your perspective of the elements in your Program, a checklist for the architect's and contractors' scopes of work and new insights into the difficulties with which both of them are confronted. A little empathy for their concerns can get you a lot of mileage in the respect area.

Database

I've mentioned the *flat-file database* several times. That's distinguishable from the *relational* database. Both of them store information so that it can be coded and accessed selectively, and sorted in useful listings.

The relational database has complex algorithms (which is computerese for internal formulas) that make it possible to code and access data from a large range of structurally unrelated information. The learning curve for using a relational database is fairly long, and we don't need it in construction.

A flat-file database can store all the information you will ever need to track the history of a construction project, to organize the elements of your Program, and to provide you with effective checklists. It's much easier to learn, and easier to code and access.

See if you can get someone to introduce you to one (there are several available for PCs) before you get too deeply into using either CPM or defining your Program.

You'll find that if you're able to establish a well-coordinated, cross-indexed system of codes, they will work for your database, your schedules and your spreadsheets. The three in combination can be a powerful resource.

23

Drawings

I could take you through the organization of the typical design drawings. You'd be bored and possibly confused before I got through two pages. It takes a lot of words to describe what you can learn quickly, hands on, with your guide's help.

The best way to get into reading drawings, and to eliminate the intimidation you will certainly feel if you haven't been trained to read them, is to sit down with your guide and those drawings and other documents in the Materials List in chapter 22.

Start into them easily. See how they're organized ("C" for civil, "A" for architectural, "S" for structural, or something close to these). Look for scale indications: 1 in. = 1 ft., but don't get hung up yet about scaling the drawings. Find an "A" plan view and an "S" plan view of the same area and overlay them on a light table, or a backlighted picture frame. Check out how the column lines are numbered (numbers in one direction, letters in the 90 degree direction). Note where the north arrow is. This should help orient most of the plan view drawings.

Now find an "E" or "M" drawing of the same area and overlay that. You're beginning to check the *coordination* of the drawings.

Next, look for a line with arrows at the ends both pointing in the same direction drawn across one of the "A" or "S" plan views. What the designer has done is to "cut" the drawings at that point, and "look" in the direction of the arrows. The arrowheads will identify the drawing number on which you can see what that view (probably an *elevation* or a *section* drawing) looks like.

You're on your way. Your guide can take you from there. Pretty soon (unless this was an exceptional project), you'll find details, or dimensions, or locations of things that don't work.

You will also start to amaze yourself as you put the drawings together with, first, your bar chart and then your CPM, what you will begin to see. And after a while the process will reverse: when you look at the CPM analysis you've drawn, you won't see squiggly lines, you'll begin to see the details those lines represent.

> When the day comes that you find yourself working with an architect who's been able to add some bells and whistles to his inventory, you may be introduced to Computer-Assisted Drafting (CAD) and Walk-Thru colored animations of your project. These are whiz-bang marketing tools for the architect. CAD

should actually make it quicker and easier to provide you with periodic checksets during design (even if some of his other consultants don't use them) and to make corrections.

The beauty of where you will be coming from, though, is that you will recognize these tools as aids. You will already be ahead of the power curve where substance counts: where the organization, the coordination, and the accuracy of your construction documents meets the road: in their compliance with your Program and your budget.

Specifications—Divisions 2 thru 16

Divisions 2 thru 16 are what are known as the "technical specifications." The numbering of each division was fixed long ago, and contains a standard listing of items it covers. Division 16 is always electrical, 15 is mechanical and so forth—although the specifics of those items within each division will be different on each project.

All you want to do at this time is to scan how they're organized, and check a few references to the drawings, or vice versa. Then, put them aside for now.

It's possible, though unusual, that there may be an extra division or two. The CSI (Construction Specifications Institute) and the industry generally did a very credible job of setting up sections of the specs many years ago. But the world has moved on apace, and occasionally there is a strange animal that doesn't have a place in the standard divisions.

Specifications—Divisions 0 and 1

This section of the specifications is equivalent, in contractual terms, to the procedures you will be defining in your Plan. It instructs the parties to the contract (owner included) what his duties are, what standards are to be met, what criteria will be used to measure those performances, how parties will be paid, how disputes will be resolved, and special considerations unique to the project they cover.

There are always two parts to these divisions, and often more.

The basic part is known as the General Conditions. These are reflected in the AIA A201 General Conditions. Many organizations have developed their own. A highway department for instance will have one unique to its organization. They're often referred to as the "boilerplate."

Because the boilerplate represents the way that organization intends to do its design and construction business, it needs to be supplemented by "Special Conditions" or "Supplemental Conditions" in order to cover the particulars of each respective project.

The one awkward part of this system is that the *supplementals* sometimes override the *generals*. Careful reading is required to make sure that there are no inconsistencies between them.

As you're working with your guide, make a point of reading the respective sections back and forth (say, about time extensions, or schedules, or payment, for example) to see how they can get bollixed up sometimes.

Putting Your Tool Kit Together

Here's where you will spend a considerable time with both your guide and your attorney once you have been through the other tools above.

You'll scan the daily reports, the sample submittals, the RFIs and the change orders your guide has accumulated, and begin to itemize where they belong in your procedures. You'll start to define with specificity under what conditions, with what promptness and frequency, and at what level of detail you will require that change orders, daily reports, meeting notes and other documentation will be delivered to you.

> Again, only as an example, you should require a daily reporting system that identifies not only how many people are on the job each day, but how many with each subcontractor, where various crews are working, and the description of the work they're performing, plus the equipment in use.

> Now, you're beginning to get data that can be used to report actual work against the work scheduled.

> For more specificity and lists for your considerations and ruminations as you refine your procedures and specifications, refer to the subject headings of WHERE HAVE WE BEEN?

You'll start to look at definitions to decide whether they adequately cover all conditions.

For example: the description "abnormal weather" doesn't mean a thing, even in a given time and place. Neither does "normal rain days." Sometimes, it can rain a hundred-year storm and have little effect on a closed-in project. Other times, one heavy rain on a weekend can muck up the site so it's unapproachable for a week.

You'll look at who gets the first word, the middle word and the last word in disputes, and make some notes to discuss with your attorney.

You'll define who will keep the progress meeting notes. It needs to be [1] someone who will have no axe to grind in the event of any dispute, including one about the transmittal and handling of paperwork. And it needs to be [2] someone who will raise meaningful issues and who will associate time values to them so that they don't delay the project.

You'll look at the scheduling specs to decide not only whether they're appropriate for your project, but who will follow up on them, who will ensure schedules are kept current; and how the schedules will be used to negotiate delays. These are complex subjects, but very important in terms both of negotiating change orders and documenting disputes and delays.

You'll want to go back to some of those chapters in which we listed the interactivities during construction (again, WHERE HAVE WE BEEN? is a useful reference for this purpose) and make sure that the requirements for the architect and the contractor (and the CM or any other representative) to log these transactions and to forward them to you regularly (weekly is appropriate) are adequately defined in your procedures, so they can be transferred, with the assistance of your attorney, into the specifications.

By this time, you're likely to be getting a pretty good handle on what you will want in the administrative specifications for construction. You're likely to have additional time (unless you go with design/build) to refine them before construction.

Design Team Specifications

Now, it's important to reflect the mirror image of all of these transactions in both the construction agreement with your architect and in his own contract to produce the construction documents.

23

You will have started some of the process by defining the requirements for the architect's design-production schedule, his budget and his payment control. Now, it's essential to incorporate the reciprocal requirements into his performance agreement for construction.

No more "promptlys" for responses to RFIs, change proposals, and submittals. You will convert the architect's performance requirements, as you have the contractor's, into something along the lines of:

> *"... respond so as to meet the requirements of the currently accepted critical-path schedule on the project."*

Owners have been led to accept the idea that the construction schedule is for the contractor to follow. That totally disregards the realities of submittals, shop drawings, ASIs, RFIs, changes, unknown site conditions, disputes, inspections, etc., etc. All involve the participation of not only the architect, but of the entire design team.

If these transactions aren't tied into the contractor's schedule, the owner has just handed the contractor a whip with which to beat the project. If the schedule isn't maintained, the whip is probably a sword—because the owner has specified a system to which the owner is patently not committed.

If the owner and architect are not committed to understand, insist upon and perform under the scheduling system specified in the contractor's agreement—the owner would have been better not to have specified his own system to begin with. Better to let the contractor do his own thing and argue about the consequences later—at least, the owner has not implicitly abandoned his own specifications.

> *Don't get me wrong. It creates a whole new set of problems not to have a competent time and cost measuring system on the project. I raised the issue above by way of suggesting what happens when the owner specifies something—anything—and then lets it slide.*

Reasonable Implementation of these Tools During Construction

Now, let's discuss how these issues of scheduling, cashflow, payment and recording of interactivities can be used by the owner reasonably and realistically.

330

We will return to the Plan in the next chapter, and emphasize principally its application to design, on the premise that this approach will gain optimum benefits for the project and for everyone working on it.

If the owner selects the right architect for his project and his Plan, then the evolution of a design-phase *partnering* relationship (I only use this misnomer because it's in current usage) should evolve as a natural enough process between the owner and his architect.

That should make the requirements for scheduling design production no less stringent, only more reciprocally comfortable.

"CPM—It Will Save You from Disputes"

The state highway department was within a few weeks of putting a $9 million bridge project out to public bid. The Director was concurrently attending a claims seminar in which he heard those words above. He promptly directed his staff to incorporate a detailed CPM specification into the contract documents.

The borrowed specification required that the contractor process the information in a specified software system. The contractor had extensive experience with the system. Neither the department nor its design consultants had prior experience with *any* CPM system on *any* construction project. The state hired an outside computer consultant to process the contractor's data.

Substantial disputes arose about design defects, unknown site conditions, inadequate surveys by the state of existing conditions, plus administrative delays by the state to resolve these problems. The state repeatedly demonstrated its inability to recognize or assess the relationships of design, problems and progress to the schedule during construction. It repeatedly refused to grant time extensions.

In the arbitration that ensued, the contractor ran circles around the state, which continued to demonstrate its inability to apply the system it had specified to the analysis of the project, and repeatedly drew inaccurate conclusions from the scheduling data.

The state was required to pay the contractor almost the full amount of its claim, in delay and acceleration damages.

But the relationship with the contractor has a different flavor. First, there are many more parties to deal with. Second, there never was a set of perfect construction documents (although getting close, within 2% - 3% of the budget, is attainable).

The contractor is unlikely to build the project just the way you laid it out in the B/BA. He will be constrained by his available personnel, his equipment, the subcontractors he has put together, and other variables.

Neither can he be expected to be any more prescient than the owner and architect have been. The schedule he prepares is likely to show cracks, if not outright flaws, within the first few weeks of construction.

One of the most damaging things an owner can do (and scores of public agencies believe they can do this with impunity) is to [1] disallow the contractor to change his schedule once he's set it down, or [2] to ding a contractor (say, by withholding money or some other Draconian measure) for missing the precise dates of a series of 3-, 5- and 8-day work items within his schedule.

No lives, no drawings, no days in the week are that precise.

What the owner needs to be concerned with is whether the contractor is on a reasonable track to meet important intermediate milestones, and whether the contractor has planned his procurement and resources with this in mind.

Most schedule specs have a requirement that specifies that the schedule be updated if a certain amount of time is lost. Too often, it is disregarded by both the contractor and the owner.

It's an important requirement. Used effectively, it avoids the unnecessary and unreasonable arguments about missing unimportant dates. But it must be used scrupulously so that none of the parties gets caught up in a post-mortem finger-pointing exercise.

It's also to avoid this kind of exercise that the logs of *hidden* activities, such as the processing of submittals and RFIs needs to be maintained regularly, too.

What the reasonable owner hopes to achieve by requiring the contractor to provide a competently detailed CPM schedule at the start of the project is [1] to demonstrate that the contractor has thought through the job and hasn't missed something big and [2] to alert the architect with regard to anticipated submittals and inspections. The owner will also need to be able to identify in this schedule his own scheduling of participation and decision-making during construction.

There is no good reason that the contractor should not be allowed to make substantial changes in his schedule during construction.

There are, however, some preconditions that need to be defined before substantial scheduling changes are made:

- Substantial changes should only be allowed if delay or other disputes, which could in any way be associated with the change, are resolved beforehand.

- The changes should be made well enough in advance of projected re-scheduling so as not to interfere with the balance of the project.

- The owner and design team should be alerted well in advance of consideration of the changes, so that the full implications can be digested before they're implemented.

A great many projects have gone into unnecessary tailspins because foolish and/or incompetent owner representatives have made big issues about small items that could be readily absorbed within a reasonably and realistically maintained progress schedule.

Project Meetings

Too many meetings are convened to hear about problems for the first time.

Following is a procedure JBA put into specifications 25 years ago that has worked with considerable success on many projects:

Monthly progress meetings are moderated by a representative of the owner who is *not* the same party retained to manage and monitor the daily activities or process the daily paperwork on the project (as, for instance, the CM typically is).

This owner representative (or OR—he or she has many of the characteristics of the guide) is positioned to maintain and/or confirm the objectivity of meeting notes, and to ensure that agendas are responsive to activities on the job.

Three to five days before the monthly meeting, the contractor prepares his proposed payment request by reporting his as-

23

sessment of percentage against cost-allocated items in the schedule. The CM does the same on his copy of the schedule. These payment proposals are exchanged, with copies sent to the OR not less than three days before the scheduled meeting.

Also exchanged and forwarded are: [1] sub-networks, or fragnets (you'll learn about them in A-2), which reflect how changes are proposed to be inserted into the schedule, [2] statements of problems by the contractor and/or CM and/or architect, [3] proposed agenda items and any special conditions which have come up, and [4] an update of the schedule, projected to the date of the meeting and incorporating the fragnets.

The OR has three days to study these reports. The participants have the same time for each other's reports. (The reports aren't formal, they're working papers, but they are in standard formats so that no one has to figure out what he's looking at.)

As a result, meetings are directed toward [1] resolving problems, and [2] resolving differences.

What's interesting about these procedures is that, even if the owner doesn't make a point of it, this is what occurs:

- After a couple of meetings everyone gets the idea that problems will be resolved quickly and realistically.

- Differences quickly become minimal. Everyone learns where everyone else is coming from.

- Meetings get shorter and everyone gets quickly back to business as usual.

- There is a substantial reduction in obstructionism, disputed items and EWOs.

To those of you who are saying—but that's what partnering's about—note:

- This process started with the revised design and construction agreements and specifications, plus explanation of the associated procedures in the original RFQs for design services.

- It is enhanced by having better construction documents.

- It was also enhanced by establishing systems that everyone follows and understands, including standardized scheduling and reporting systems. There was no wasted motion trying to read unintelligible documents. Everyone was reading from the same page right from the start of design.

Who Owns the "Float"?

You will hear a great deal about "who owns the float" on CPM scheduled projects. Most people think that the answer is "whoever gets there first." So the owner claims the float for itself (public agencies are notorious, again) and implements a major change order, agreeing to pay only the direct costs thereof.

Some not-too-well reasoned legal arguments have been put forth to support this view. The result has been that contractors take a lead from Parkinson: They fill the time available to minimize the float sloshing around in their schedules.

This action on the part of the owner to grab the float may—or may not—entirely disregard that a change, even if it's not on the critical path, can impose other disruptive conditions on a project.

It is a reasonable consideration for the owner to keep in mind that the contractor may have some inherent right to the float before the owner claims it. It can be a resource he has considered in the context that he may have to shift resources to keep up with the critical-path work if he is delayed because, say, a supplier defaults. Much depends on the reasonableness and mutual acceptability of the original schedule (See comments on this issue in chapter 14).

The point of these discussions is that reasonableness needs to be the keystone when people sit down to review a schedule on a construction project.

Reducing Risk

The focus of this book is to expose the pitfalls and the risks that await the unwary in the design and construction of building projects. Its purpose is to assist in reducing those risks.

The outline and the curriculum that follow are suggested to assist a committed member of an owner's organization to develop a Project Plan that will be productive for this purpose.

Float

To get a handle on the idea of "float," go first to the definition in the glossary. Then you will need to spend some time with Appendix A-2.

The question in this heading is an important one because "float" is a resource that everyone on a construction project wants to have in his or her own pocket.

335

23

Before proceeding to a curriculum for developing an understanding of the tools, it might be useful to recap where we've been:

> The owner's purpose is to produce a successful project: one on which design-related changes do not exceed 2% to 3%.
>
> An equally important purpose will be to produce a project that is not exposed to litigation and to eliminate, if possible, even the potential that a claim will be filed.
>
> As an ancillary to these definitions of success, the project should make it possible for all the participants to optimize the use of their own resources.
>
> In order to accomplish this, an owner must first become familiar with the territories, the interrelationships and the processes that are inherent in the development of a construction project.
>
> Having developed a familiarity with the processes, the owner can then proceed to take the lead in developing procedures and contracts that will facilitate these definitions of success.
>
> These procedures will incorporate enhanced means for selection of those who will design, construct and manage the project; and enhanced means for measuring and documenting their performances.
>
> Because the owner initially will be unfamiliar with these processes; and will likely have limited time to develop an understanding of how they work, it will be useful for the owner to find a guide who can facilitate and accelerate his or her learning process.
>
> As the owner acquires these enhanced perspectives, they will come together into a definitive Project Plan.

Even now, before you have developed your Plan, you should know considerably more than you did before about the strangers with whom you'll be working, and about the ones you will want to avoid.

You may already know more about the overall processes of design and construction than many of those strangers do. You should have some new perspectives about yourself and your organization, or be on the way to finding out.

What remains may appear to be a daunting task, so it will be useful to consider what you won't be doing, together what you will need to do.

CONSTRUCTION

You won't

- Be assuming the role of manager of the construction contract.

You will

- As Leader with a clear Plan in hand, have defined clear performance standards and scopes of work for whomever you have selected to represent you during construction. You will have become comfortable with the documentation those representatives will be directed to maintain on your behalf.

You won't

- Be trying to take—or allow your representative to take— initiatives away from the contractor that have been provided to him by his Contract.

You will

- Be asking—either directly or through your representatives—the contractor insightful questions that will allow you to stay ahead of problems. These questions will inform the contractors that you know where you're going, and that you have at least some understanding of the problems they may be encountering. You will be advising them that, to the extent they're not contractor-caused, you and your representatives are prepared to assist in any reasonable way as long as it does not present a liability for your budget.

DESIGN

You won't

- Be accepting standard performance agreements out of hand, or agreeing to payment schedules without adequately defined production.

You will

- Have defined (in advance of even interviewing design consultants), requests for qualifications and for proposals that have made it clear they will be working under procedures and within a Plan through which the owner intends to be fully informed. You will have defined a

23

Plan for progressively confirming the coordination of the design and contract documents and their compliance with your programming and budgetary limits.

THE LAW

You won't

- Ask your attorney of choice to "look over and bless" some standard contracts.

You will

Since there is no contract and there are no tools that are fail-safe to protect the owner against problems or construction claims, the absolutely best resources an owner can establish on a construction project are reasonableness and respect.

- Have spent the time necessary for you and your guide to have defined what you believe you and your organization can effectively administer through an explicit system of procedures. You will direct your attorney to work as one of a three-member team to define integrated design- and construction-contract requirements to implement these procedures.

You won't

- Be misled by the cloak of intimidation in which the law shrouds itself.

You will

- Demand the services of an attorney who will get down and dirty with you and your guide to talk practical sense about how to improve the performances of those you hire, plus how to sell your consultants' sureties on a program that will protect their interests as well as yours. Or in the alternative, find another means to protect yourself against design liabilities.

It's important to say one more time:

Since there is no contract and there are no tools that are fail-safe to protect the owner against problems or construction claims, the best resource an owner can acquire on a construction project is respect.

Respect comes to those who have schooled themselves in how the interrelationships work on a project, and who function with reasonableness and integrity.

A Study Plan

I don't know how much time you have before you undertake your next project. Or your level of familiarity with the processes and the tools you'll need.

As a point of reference, we'll start with a study plan as though you have three months before you intend to interview architects. If you have more time than this, most of it should go into Phase 3. That's where you will need all the time you can provide.

If you have six months or more, you might consider finding other owners in your position and asking them to join you. The interactions and added insights could be useful. (See page 314 on the possibility of forming an Owners' Association.)

However you approach Phases 1 and 2, you will have to maintain a rigorous schedule to get through Phase 3. You, your guide and your attorney will need to set up a series of what-if sessions to refine the procedures, specifications and agreements you intend to develop. You will need to invest considerable time and effort to get them accepted by your board, your agency or your organization. Phase 3 will be the first test of the planning and scheduling tools you'll pick up in Phases 1 and 2.

A 15-Week Curriculum

With only 15 weeks before you need to start design, I suggest you divide your time this way:

Phase One: Weeks 1-3

- Find your guide.

- Get comfortable with your tools. Spend one or two evenings a week with them. Then, spend a couple of half-days with your guide in which you talk out possibilities, and you get some of your questions answered, or ask him/her to undertake some research for you. Maybe he/she can join with a family group to schedule a vacation. You're likely to learn a lot that you'll carry down the road with you.

- You should spend a couple of evenings over a warm computer, experimenting with formulas, assigning codes to data, and selecting and printing reports.

Phase Two: Weeks 4-8

- Get serious about outlining the procedures you intend to put into your design contract (see WHERE HAVE WE BEEN?).

- Set up the criteria you want to find in your attorney—or if you already have one, a litany of questions and issues you want to discuss with him/her. In particular, you want to learn whether you will have a constructive ally or a naysayer to deal with. You don't need the latter.

- If you have found him or her, it's time to bring your two closest allies together and to get down and dirty about how good or bad the AIA documents are for your purposes, and whether either your guide or your attorney has a model to start with that won't require reinventing the wheel.

- Develop a series of definitive checklists that itemize the documents you will need to produce in order to get your project under way (Design RFQs, interview procedures and evaluation criteria, back-check procedures and questions, design contracts, your Program, outline of construction procedures, etc.).

- Use your newly acquired CPM skills to schedule the scopes of work you have parceled out for you, your guide and your attorney.

Phase Three: Weeks 9-15

You'll need every bit of this time:

- To formalize, print, review and discuss your procedures and documentation system.

- To prepare, confirm and print contracts.

- To meet with potential sureties and modify your requirements, if necessary. (You're trying to get ahead of the objections you'll hear from architects.)

- To prepare the sales and marketing proposal to sell your Plan to your own organization, to meet with them, and to satisfy their objections. In this regard, see chapter 25.

- And to confirm the details in your preliminary Program.

Your developing facility with CPM will be an important resource to keep Phase Three on track.

24

YOUR PROJECT PLAN

"First, set mind at peace..."

As the author of *Zen and the Art of Motorcycle Maintenance* stands in a garage waiting for service, his head pummeled with the noises from a boom box, he opens a maintenance manual that starts out *"First, set mind at peace...."*

I have a rock, overlooking a valley, where I sit when I need to sort things out. The only sound is the breeze in the trees. Sometimes, I bring Beethoven's *Ode to Joy* with me.

It may become important for you, too, to find a place of peace and inspiration as you set out to put new tools into your toolkit, and to design a Plan that you may have to fight for. Since I found my rock, the world has acquired a lot of new sounds, though, so I don't want to suggest there's only one kind of environment in which to consider what follows in the remainder of this book.

But it's important to suggest that if you're going to organize into a workable Plan the questions in your notebook, the answers you'll get from your guide and attorney, and all the territories we've passed through, you're likely to need your own special corner to sort things out from time to time. Putting unconventional ideas together into a workable scheme isn't the kind of exercise that mixes well with the rattle of other daily activities.

Designing and implementing a Plan that runs counter to conventional ways of doing things will require you to be on close speaking terms with your convictions. You will want to have walked through a considerable number of *what ifs* before you're tested under field conditions.

Preface to Developing Your Plan and Program

While you're putting together the *how* of your Project Plan— you'll also be preparing your preliminary Program. You may find some of the following thoughts useful.

> If you're renovating a building you can't give too much consideration to surveying existing considerations: utilities, available power, condition of equipment, compatibility of on-site

341

materials and equipment with those that may need to be added. Exploring behind walls, in crawl and overhead spaces. Checking the condition of the roof. Sometimes the extent of improvements will trigger requirements for compliance with current codes you may have thought do not apply. Your architect should have all these considerations incorporated into his scope of work. Your guide may be able to provide some insightful recommendations.

Whether you're building on a new site, or one which formerly held another building, you will certainly be required to undertake environmental studies. But whatever the environmental requirements, almost no one does enough subsurface exploration of a virgin site. You will need to consider how you expect to share risk for unknown site conditions. The simplistic contract terms that many owners have used to try to transfer all risk to the contractor have often been overturned, or ineffective.

> Despite the absurdity of the premise, there are still public agencies, for instance, that continue to believe the clause that *the contractor is required to inspect the site* will insulate the owner against any unknown site conditions.

If your project is a school, you'll be bound by an arcane body of rubrics lobbied into the regulations of your department of education. But the stories scattered through this book should suggest that there are reasonable and effective alternatives to paying 8%, 9% or 12% to an architect for the design of underground utilities, paving or electrical conduit. Again, a knowledgeable guide should be able to discuss these and other options for optimizing the limited funds available for your school program.

If you're a hospital administrator, university director, or business organization with an ongoing building program, the administration of your projects may be under the direction of an in-house architect or someone who attained the position because of his success in real estate or some other non-construction related enterprise. Or your construction program may be beholden to the accounting or legal department. If you or your board believe there are potential benefits for your project in the ideas here, you may find it necessary to consider some modifications in your organizational structure before any significant enhancements can be made to your contracts, your development procedures or your administration of design.

The Heart of Your Plan: Some Scenarios to Consider

The heart of your Plan is the Buildability/Budget Analysis. To make it work, you'll need the cooperation of your design consultants.

To ensure that cooperation in advance, you'll need to make sure the entire design team is aboard before you hire your architect. As with subcontractors or suppliers, it only takes the deficient performance of one among the design consultants to mess things up.

Your Requests for Qualifications from your design consultants will have to be very explicit. You will want information related to all the projects they and the architect have worked on in the past 5 to 10 years. Names of owners and contractors, not a selective list of owners only. After you have decided which design team rates highest, you will want to know more about how the members of that team performed on a wide spectrum of projects, and responded when the inevitable problems arose.

You will want to be very explicit about how you intend to require definitive production schedules and buildability reviews throughout design. Those who object need not apply.

You will want your RFQ and those procedures disseminated to everyone the architect identifies as on his team. The principals of major disciplines should be present at your interviews and when you confirm your design contract.

That requirement alone is likely to discourage firms whose offices are at a considerable distance from your offices or jobsite, since you will also be notifying them that you intend to meet several times during design to review your analysis of the documents with them.

The trade-off for you, if you retain a design team whose diverse offices are scattered or at a considerable distance from your architect's, is that you will have the all-too-prevalent experience of not being able to follow the quality of your construction documents, until there is insufficient time to do anything about them before they go to construction.

343

The Long-Distance Architect

The owner was building the flagship of its City Center: a multi-purpose auditorium. The agency was determined to retain an architect with a national reputation. Despite the fact the project was in the southwest, they retained the New York architectural firm that fired their collective imaginations with a dramatic theme. There was no discussion of logistics, of management expertise, nor of how the architect would assemble its design team.

When the competitively bid, lump sum contractor started to raise issues of design ambiguities and of the difficulties in getting timely resolutions, the owner started to learn how widely spread the design team was: a structural engineer in Michigan; a mechanical consultant in Atlanta; the electrical consultant just down the road; and an executive architect (to handle local problems) about 400 miles away. The principal whose name the architectural firm still operated under had, in fact, been retired in France for the past two years. He only participated in interviews.

The disaster that ensued added 27% to the cost of the facility, and required 3-1/2 years to resolve in various arbitrations, with the aggregate expenditure, all-round, of an additional $780,000 in legal fees.

When you do interview your design teams, you and your guide will be equipped with some challenging questions and a system for evaluation more definitive than the one most boards, committees and agencies use—the 7-line-item scoring sheet with a possible score of 100. (WHERE HAVE WE BEEN? will assist.)

As you consider these items, you will also need to be thinking about what you will require as a standard you believe should trigger the architect's liability in the event of design-generated change orders during construction.

You've already read about wasting policies, so you know you'll have your attorney check into that possibility. But you may take a more constructive approach. You know, and so does your architect's E&O carrier, that the B/BA Plan you intend to implement will reduce your architect's and their potentials for disputes, too.

If you're willing to make an investment in the quality of your documents, by not only checking their quality, but by being more responsive than many owners are to the need for decisions during the design process, the architect's carrier should be willing to make a realistic adjustment in the coverages your architect provides on your project.

You might even investigate the alternative of insuring against design-caused claims with litigation insurance.

At the very least, you, your guide and your attorney will need to have a very serious discussion about how much design error protection you need, and about how you'll get it..

Your Plan: Checklists and Checksets

As we came through the various territories in this book, we generated listings suggesting the qualities you will want to identify in the various participants on your project, lists of interactivities among those participants, and questions you will want to ask.

You will find the checklists of what you expect to receive in each design phase in the earlier chapters. To facilitate your reference to these lists and descriptions see WHERE HAVE WE BEEN?

The architect's schedule and agreement will define points in time at which you will receive [1] a checkset which you can use to develop your own independent review of schedule, buildability and costs; and [2] a billing based upon the labor hours and costs expended to that date. Payment to the design team will depend on how closely both of these documents (the checkset and the invoice) relate to the architects' schedule and budget. You might consider a retention system to ensure compliance.

Your review should result in a meeting with all the major disciplines of the design team within a reasonable time (a week to 10 days) after you receive the checkset documents. At these meetings, you will review your findings and provide the team with questions and or requirements to maintain the design on schedule, consistent with your Program, and with standards for buildability.

Using the B/BA

As you consider how you plan to use the Buildability/Budget Analysis to monitor and manage the design process, there is an important consideration to keep in mind:

24

The B/BA is intended not just to protect your project against delays, construction claims and legal costs. It is also intended to help you avoid the kinds of budget overruns that will ruin your day.

You will be integrating into your buildability-scheduling analysis a level of cost allocation to refine and confirm your budget and to establish appropriate contingencies. (See chapter 17.)

The product of each of these B/BAs will be a checklist which you expect the design team to take into account as design progresses.

The final product of your B/BA at the completion of the CDs will be a collection of schedules, checklists and budget/cost allocation that you will keep in your files.

It is not the purpose of your B/BA process to define how the contractor should construct your project. In case no one has told you: that is one of the mortal sins in construction. The contractor has full discretion to build your project in any way that is consistent with the design and the constraints you put into his contract. If you have any constraints to put on the contractor's performance, make sure they're reasonable.

Your B/BA will tell you a great deal about the responsiveness of bids. It will be valuable when you check the CPM schedule and cost allocations your contractor provides for their reasonableness. It should help you prepare your bid documents. It can be useful in the negotiation of changed conditions and contract modifications.

But its principal purpose is to shake as much water as possible out of the architect's documents, and to make sure they meet the standards we've repeated so many times already.

How You Plan To Do Project Business

When you're ready to invite consultants for interviews, you need to be ready with a clear game plan.

You don't want to do what most owners do: present your interviewees with the diverse concerns and personalities of a "committee," or the prospect of "management by committee."

A Certain Flexibility

Once you develop the outline of procedures for your Plan to measure design, you'll want to refer to the other types of documents you'll receive during construction.

Every project is different. Every contractor has his own set of reports. As long as their content meets the criteria you've established in your Plan for adequate reporting and documentation, then it makes sense, and it will contribute to the overall atmosphere of cooperation on your project, to accept the contractor's forms when you can. You're not trying to reinvent a wheel that already turns efficiently.

Neither are you trying to change the entire industry. Your architect and your contractor may be doing things in different ways on other projects. That shouldn't concern you.

Once you, your guide and your attorney have set out a clear Plan for what you require on your project, the important thing is to stick to it, but to be reasonable in accepting the information you require in a format either your architect or your contractor is already using *provided it doesn't compromise your purpose to have comprehensive documentation of their performances.*

By demonstrating right up front that you have a clear Plan, and that you're prepared to provide clear directions and decisions, you will get entirely different reactions and responses from those you interview. Your specific purpose is to challenge and confirm the competence of those from whom you're entertaining proposals by demonstrating the directness and candor with which you expect to administer their performances.

You, your guide and your attorney will have had to assure your district board, your assembled citizens or your reluctant associates in advance that it's not your intention to subvert their prerogatives, but that your purpose is to bring them a design team that understands how your organization intends to do business, and that your organization has clearly defined the rules of the road.

These preparations for how you plan to administer the performance of those you hire are not intended to obviate the need for your internal committees and reviews. You may have to explain that what these representations really means is that your Project leadership team—the three of you—have shouldered the responsibility to formulate the questions and accumulate the information your board or committee will need to make its decisions.

If you think about it, once the contracts are executed, it's you, the leader, who will take the lumps if things go wrong, anyway. The "committee" and the board will have disappeared back into their daily lives, to surface only to bless periodic reports or to share their displeasure with you.

If you want reality on your project, it has to start with the leader leading it.

The consultants and contractors you retain will appreciate it. They will know that they're working with an informed and intelligent owner who took the trouble to learn about their business and their concerns, one who has a Plan for running the project. The message they'll get is that they can expect timely answers and effective action when it's required during the project.

By the time you get to construction, contractors will also have gotten the word (even without you advertising it) and will expect to find the kind of well-organized documents and procedures they want to bid and work on.

Believe it or not, both the design consultants (despite some potential early grousing) and contractors will heave a sigh of relief that they're not once again working with an owner whose hand they will constantly have to hold, or who will be flopping like a fish whenever a timely decision has to be made.

If you have an organization behind you, with resources to complement your ideas, to fill in the blank spaces from a reservoir of past experiences, you're well on your way to setting up the project team you want in place to help you select design consultants and the form of construction contract you believe will meet your requirements. And you may be ready to develop and refine the contract requirements to carry out your Project Plan.

If you don't have those resources, or if your experience with your present associates tells you that you need more energetic, objective assistance than they can provide, it's time to look for one or two special associates from within your organization who will join as allies to assist you with "selling" your ideas and your Plan.

That Other 20%

The two most frequently asked questions about this book have been:

Why build it twice? Won't I have enough trouble building it the first time?

Why can't I expect to avoid 100% of the problems of construction?

You now have the answer to the first question. But no matter how detailed your B/BA, there will never be a construction project that will not encounter the unexpected.

There is a place where the two ideas meet, however. The more effort you put into analyzing your project through the B/BA, the better you will be prepared to respond to the unexpected, which should at least reduce your exposure to the costs the unexpected can present.

To help your thinking in this direction, even as you develop your Project Plan, here's a short list of some of the things that can come up on a construction project. (You should add to it as you evolve your Plan.)

- The as-built documents you inherit are inaccurate or deficient. You contractor will encounter hidden conditions you didn't show in your documents; or that are shown inaccurately. (One way to reduce this exposure is to undertake extensive pre-design exploratory surveys and destructive testing.)

- The surveyor's work leaves you, your architect and your contractor with locational errors.

- Soils and underground conditions are substantially different from what anyone expected.

- The building department, the inspector, the fire marshal or other agency representatives impose entirely arbitrary interpretations on your project after you've committed it to construction. Or one of the review agencies takes longer than you expected to approve your documents, or change orders that are implemented during construction.

- Strikes, delivery delays, weather or defective equipment impose unexpected delays on performances. (No matter that some of these delays are "non-compensable," they will impact the owner's budget.)

- A major contractor or supplier goes out of business. (However the cost implications are resolved, this will have major impacts on the coordination of the Work.)

The Architect's 21% Contingency

The project went to bid with a typical 10% contingency in the budget. Under competitive bidding, the project was awarded to the low-bid contractor for just 2% over the estimated costs. That left 8% in the contingency.

In the first three weeks of the project, the contractor forwarded 16 RFIs to the architect. Several of them reflected dimensional errors, ambiguities and omissions in the documents. The architect sat on them.

Then, the owner requested a change—one that would have consumed about 3% of the contingency. The architect forwarded the change request to the contractor under "Owner Requested Changes" and incorporated several of the issues raised by the design errors. The change orders came back adding a total—8%—that used up the rest of the contingency.

The architect recommended approval by the owner, who executed the change order. The change order buried the 5% overruns that were the result of design errors and omissions.

More RFIs followed. The architect started to incorporate credit items with successive changes to compensate for errors, downgrading the quality of plumbing fixtures, hardware, paint schedules, floor coverings and some equipment. All were forwarded as "owner requested changes."

By the end of the project, a total of 21% of design errors and corrections had been offset by the balance in the original contingency and reductions to the quality of the project.

The owner, unaware of the extent of the changes in quality because of the adroit way in which the architect had buried them in successive change orders reported to the board that, *"The project was completed within budget."*

The architect went on to promote to other owners, as it had to this owner, that it had a record of *"consistently delivering projects within their budgets."*

The architect's practice of burying his errors in "owner requested changes" only became evident when the owner's aware operations engineer, concerned about the quality of some equipment, started to back-check the original specifications.

25

SELLING YOUR PLAN

You've Got to Be Believed to Be Heard

You're convinced. You, your guide and your attorney have worked out a Plan.

Now it's time to sell it.

Time for some dry runs. There are others you need to bring aboard.

The first are the other members of your board, your agency, your upper management, your staff.

With the commitment of your organization in place, the next to be convinced of your sincerity will be the consultants you intend to hire. Some of them will not be prepared to provide you with definitive information, to be measured, documented, checked and balanced. You are likely to lose some of them, and that will become a challenge to your commitment.

Once you select your design team your sincerity and commitment will need to be consistent through negotiations, meetings and in requiring responses to your B/BA reviews.

Your face and body language will convey plenty about your convictions.

It may be at times like this that you will need that rock, that quiet place to reconstitute your commitment. You may as well get used to being alone sometimes. Leaders often are. So it's essential that you take these convictions and settle them down inside in a secure place where you can find them when you need them.

It takes solo time to do these things. We're all surrounded by too many distractions, distractions that could interfere with clear decision-making. We *think* we know what we think, but finding out when the chips are down can often be too late. It's better to find out when no one else is around (unless it's your guide, your attorney or a close, confidential associate).

While you're testing your convictions it could be useful to have that comment of Sam Johnson's nearby: *"We are inclined to be-*

The Pariah

Before you try to sell anything to anybody—the architect, his sureties, the contractor, or your own organization—you might give a thought to what it means to be a pariah.

The owner who, having learned something about the hazards of project development, begins to think of himself as an expert will destroy the potential for any effective relationships with those on his or her project. He or she will be treated as a fool—justifiably—by everyone he comes in contact with for the development of his project.

Construction is a world with too many unknowns, strangers and unexpected conditions to believe that after you've developed an effective Plan, you know whatever you need to know in order to tell everyone what they're supposed to do. I've written this book after 40+ years in the industry, and I'm still surprised at how many ways the game can be played. That's why I've emphasized *respect and motivation* instead of *demand*.

Before you try to influence anybody else's activities, you could benefit from adding another aphorism to Sam Johnson's comment about gullibility and deception:

One excellent way to get anyone to tell you whatever they know, or think they know, is to share your ignorance freely with them.

It's entirely possible to make effective use of all of the ideas in this book, to develop your Program, and to put a Project Plan in place, by asking the right questions, of the right people, at the right time.

lieve those whom we do not know because they have never deceived us."

It's time to digest the chemical reactions you and others will have when you integrate the questions and answers in your project notebook with the organizations for which you expect to build that next project.

The suggestion of a mirror is a reflection (no pun intended) of sessions spent with Bert Decker, the communications guru[1].

[1] *Bert Decker has put some useful ideas along this line in his book:* You've Got To Be Believed To Be Heard.

Bert puts a mixed bag of 30 or so reluctant people in front of video cameras. They compete, separately and in teams, extemporaneously, to sell ideas to the rest of the group. These sessions are initially embarrassing to some, but they quickly convert a bunch of awkward, nose-scratching mannequins into people who, if they don't become public speakers, at least become comfortable with themselves in front of a group. They learn how to convince others of their sincerity.

You may have to do the same.

Maybe your family or a friend can get behind the video camera while you try to convince them of the ideas you'll be selling to your peers. It actually helps to have some naysayers in some of these sessions.

The dry runs about what *might* happen during design or construction are equally important. Your guide can be a productive resource for them.

You'll have plenty of opposition once you put your Project Plan in motion.

The first will come from your own organization, unless of course you have complete authority already.

Even then, however, you'll have to gather your forces as you bring together your attorney, the architect you retain and his E&O carrier to hammer out an agreement that meets your criteria for measurement of construction-document production.

You'll have to do it again as you incorporate your requirements into a construction contract, or hire a project representative to watch the store. (Depending upon the size and characteristics of your project, you and your guide may already have worked this out. But otherwise, you may still have to find a full-time representative that will meet the standards you've established.)

Whatever your Plan, it's time to do what all effective planners do: set up your guide as the alter-ego of the opposition you expect to encounter and plan the direction of your marketing program to *be ahead,* instead of behind, the power curve when others start throwing obstacles your way.

One at a Time

One more suggestion, then I'll be on my way.

This one is the result of many frustrating experiences.

Your organizational shrink will tell you that you have to consult with everybody and get their consensus before proceeding with new ideas. Do that in the area of project development, and it'll be business as usual for the next millennium.

There are too many territories to protect. Fences will go up around some of them faster than you can blink.

Over and over again, I've advised owners of organizations with continuing building programs that they should not expect success if they try to implement all these ideas at once, with all staff, on all projects, all at the same time.

When these owners ignored this caution (which happened more than 50% of the time), failure was inevitable. There are too many entrenched forces in every organization to turn them all around. There are always one or two people who feel threatened. That's all it takes.

Those two people will raise such a ruckus that their anxieties will quickly spread like a cancer through the organization. Your ideas will be in shreds within 3 hours of their presentation.

The story on the opposite page has been repeated many times. Management is convinced of the benefits of a specific new direction. They want to implement all things at once.

> They forget they're dealing with people who may find change uncomfortable and daunting. Who may find exposure to criticism—even if it's justified—intimidating.

> They forget about those who will resist—and try to scuttle—change, because they're determined just to wait out retirement.

> They don't give sufficient consideration to the different learning curves of their staffs.

> And they give insufficient attention to those who may have legitimate concerns about a too-fast implementation of new ideas.

Two Sour Apples

The organization's building program covered half of the U.S. The Vice President for development was convinced that substantial improvements could be made in the way projects were designed, budgeted, contracted and administered. He bought into the ideas in this book.

Ostensibly, too, he accepted the condition that the procedures to support these ideas should be phased in over time, by demonstrating their values through selected management teams on selected projects. Since projects were widely spread geographically, there was an ideal opportunity for phased introduction and refinement of procedures.

A meeting was scheduled at which the V.P. and I would introduce the potential program to one or two select project teams. In the space of three days, however, the ground rules somehow changed, without notice.

When I arrived I was informed that we would be making a presentation to an auditorium in which were assembled all of the organization's project and contract managers. I argued against the reasonableness of this arrangement, but the V.P. (obviously under pressure from other departments) insisted.

The presentation took only an hour to self-destruct.

Once the ideas had been introduced, two staid, prosaic naysayers alternatively had their piece of me and of the concepts that management wanted to introduce. Their objections came down to, *"What's wrong with business as usual?"* From that point, it was no great shakes to get the rest of the assemblage on their bandwagon.

As I shook hands with the V.P and departed for more enlightened pastures, I learned that the two naysayers were the in-house comptrollers who had developed 20 years earlier, the procedures and contracts with which the organization was operating.

Effective management has learned that change in many environments has to be introduced gradually. On one project with one small staff, or part of a staff that has already demonstrated an interest in improvement with selected improvements, phased in over time.

When management has followed the recommendation for phased introduction, one improvement has led to another. One team has conveyed its enthusiasms to another team.

One project led to two, two to four, and over a couple of years to a vastly improved program for contracts administration.

In an organization with ongoing building programs, there is only one way for new construction-administration ideas to take root:

- Take one project with a likelihood of success.

- Assign to it the most productive and forward thinking of your staff.

- Give the project all the quiet support you can muster and accumulate records of its performance.

- Once that performance has established enough data to demonstrate that the new ideas work, begin to publish the results (not necessarily the ways and means).

- Use the staff of that project as disciples to bring a second team and project into the program.

The Mayor's *Advances*

There was a lot at stake. A federal judge had informed the mayor that unless the city got control of its wastewater-expansion program, the government would withdraw the support of federal funds.

The mayor brought in a team with a track record of effective management on major design and construction projects. The team put systems in place that, over the first year of a 3-1/2 year program, demonstrated their value by substantially reducing costs and time on projects where they were effectively applied.

But within that first year considerable resistance had developed among the troops. Change was uncomfortable for many, especially the old-line managers. The judge returned and his review resulted in a repeat ultimatum: the city was not bringing its engineering staff up to speed, and the withdrawal of funds was still in the offing.

The mayor, faced with the choice of an ultimatum and the resistance of staff *that could not be fired,* took the one practical action that would work. Non-responsive personnel were "advanced" to non-sensitive management positions. Lower managers who had demonstrated their responsiveness were elevated to positions of authority.

Two years later, the entire staff had gone through extensive training in management techniques and extensive conversion of perspectives (with some further shifting of personnel). By the time the outsourced management team left the program, project-cost overruns had been brought down, across the board, by two-thirds.

- Within a reasonable time those who don't come aboard can go elsewhere, or be relegated to lesser projects (for the skeptical: see the evidence on page 356).

Even If You're the Leader of an Organization: Patience

Then your next moves will depend upon:

- How close you sit to the cat-bird seat in your organization; and

- Whether, by influencing, cajoling or reasoning you can convince a sufficient number in your organization that the options you want to consider will bring benefits to your organization.

Which brings us back to *patience.*

A possible variant on this theme is to implement only *some* of these new ideas on the first project. You can usually implement some of your tools for measurement of time and/or money without wholesale redevelopment of the specifications. What you need to do is at least imply a requirement for enhanced measurement of these resources in your partially modified contracts.

Then, as the benefits start to accrue, you can expand not only the application of your long-range Plan to follow-on projects, you can enhance the contract requirements and specifications as you go.

The important thing is to be patient, and expect that it will take not less than two years, possibly more, to integrate your full Plan in an organization with an ongoing building program.

This recommendation goes away, of course, if yours is a one-time, stand-alone project. In which case, prepare to link arms with your guide and attorney and to gather your resources for a flight into some interesting new territories.

There's a lot more that can be said about both successful and unsuccessful projects. Someday I may expand on the stories and some of the chapters here. For now, though, I think I've done all I should to tickle your equanimity and to challenge your ideas of what it takes to produce a successful building project.

So I'll just wish you all the best with your next one.

WHERE HAVE WE BEEN?

Grouping the duties and the attributes of the participants, the procedures and the resources you will need to develop and to implement your Plan.

WHERE HAVE WE BEEN?

As promised in WHERE ARE WE GOING?, here's an index that will assist you to develop your Project Plan and Program.

Attorney—profile, capabilities and the law

Guide—profile, capabilities and relationship to the owner

Management—definition and description

The owner's Program and Budget

The owner's Plan

Contracts

Phases of Project Development

Procedures

Interactivities and Documentation

APPENDICES

A-1
GLOSSARY

Term/Phrase	Description/discussion
A/E	Architect/Engineer: usually refers to the design team working under the architect performing design services on a project. (See chapters 8 & 9.)
ADA	Americans with Disabilities Act. A federal law whose application to a building project can contribute substantial design considerations to the owner's Program.
ADR	Alternative Disputes Resolution. The term has arisen as more alternatives to arbitration have been added to the venues for disputes resolution. Broadly, ADR refers to any legal venue for the resolution of a construction dispute other than full term litigation.
AGC	Associated General Contractors of America. An organization of which many general contractors are voluntary members. Its charter is to enhance the conditions under which general contractors perform construction work.
AIA	American Institute of Architects. The association to which many design consultants voluntarily belong, established in 1855 to enhance the conditions under which architects perform their services.
Acceleration	The use of additional resources or additional working hours to improve or to recover performance goals. Can be directed (authorized) or "constructive"— required by contractual conditions. (See "constructive" hereafter.)
Act of God	A term long established in the construction industry as a shorthand for an event or condition which is clearly recognized as beyond any human capacity to anticipate or control.
Activity	An item of work defined by trade, location, duration, etc., and incorporated into a CPM schedule. Sometimes called a "work item." (See chapter 14 and A-2.)
Addendum	A modification to the construction documents issued by the owner prior to receipt of bids and award of the construction contract. (See Award hereafter.) Contractor(s) are required to incorporate the requirements of addenda in their quotations. Compare: "change order."
Admissibility	As in *admissible evidence*. The evidence is of such a character that the court will allow it to be introduced at trial. There are extensive legal rules that apply to the admissibility of evidence. In the context of a contract dispute it means, broadly, that the facts or events in the documents to be admitted were recorded in the course of those events by someone who was a direct observer of or a participant in those events.

A-1

Advocacy   *Black's Dictionary* says: "The act of pleading for, supporting or recommending active espousal." In practice, this means that the advocate (the attorney) is not only allowed, but, in a sense, required to use every legal weapon to advance his client's position. That's why the environment of the law is so uncomfortable for the rest of us, and why reasoned negotiation offers so many advantages.

Aftermath   An apt idea to consider at the beginning of your project. If the mathematics (dimensions, budgets, quantities, markups, durations) aren't right, there is likely to be too much "after": costs, changes, delays, frustration, claims, attorneys, other strangers.... (see page 141).

Agreement   A contract executed between parties for the performance of services. The Agreement refers to the signatory document which defines the other documents (drawings, specifications, etc.) which make up the entire Contract.

Alternates   "Add" or "Deduct" options (e.g., "add alternates") associated with construction bids, which allow the owner to expand or contract the scope of the basic contract, based upon how that scope, its cost and the costs of alternates fit into the owner's budget.

Approval (Acceptance)   At one time contractor submittals and schedules required "approval" by the architect before the contractor could incorporate the work they represented into a project. The liabilities assertedly associated with "approval" became too heavy for designers, their E&O carriers and their clients. The architect now "accepts" submittals on behalf of the owner without necessarily approving what is submitted. It is essentially the responsibility of the contractor to ensure that submittals comply with the design intent of the construction documents. This, however, can lead to some difficulties if the documents themselves are ambiguous, or contain conflicts.

Arbitration   A legal arena in which construction disputes are heard, and decisions rendered, by a panel typically of one to three persons. See extended discussion in chapter 13.

Architect   The lead professional consultant whose duty it typically is to design the overall concepts and elements of a building project, and to coordinate the services of the other design consultants whose design elements will also be part of the project. See extended discussion in chapters 8 and 9.

As-Built Drawings   See "Record Documents."

Audit   Scrutiny of records. The term may be either used as a noun or a verb. It may apply to records of continuing work (usually cost and performance reports), or to company records and to completed work (often accounting records). Distinctly different perspectives and capabilities apply to either application.

Award of Contract   Notice to a contractor by the owner that the owner intends to award the Contract. The notice is often issued in advance of the execution of the Contract. This provides the contractor the opportunity to gear up when, as with government contracts, it may take awhile to obtain formal authorization to proceed.

370

GLOSSARY

Background Drawings	Drawings prepared by the architect, locating and dimensioning the architectural and outline elements of the project in relation to which the other design disciplines are required to design, integrate, and coordinate their design work.
Bid Documents	The construction documents provided to the bidder(s) from which he/they will prepare bills of materials and quotations for construction.
Bluelines	The format in which most construction drawings are printed and assembled as part of the construction documents. The word "bluelines" reflects the fact that reproducible drawings have been transferred through a process that burns away the medium on blueprint paper except where the lines are on the transparency. Many years ago, the process burned away the medium only *where the lines were,* resulting in *blueprints. (The lines were white on a blue background.)*
Boilerplate	That part of the Specifications which defines the administrative requirements associated with the construction documents. Typically, Divisions 0 and 1 incorporate the boilerplate elements of the specifications.
Bond(s) [1]	Refers to the funding mechanism used, principally on public projects, for funding the Work of the project. Example: bonds that will be sold through financial markets to generate the funds to construct the project(s).
Bond(s) [2]	The surety contract which guarantees the performance of a particular party, and provides a fund against which the party in receipt of the bond can demand reimbursement of costs if the performance is not met.
Breach of Contract	Failure, without legal excuse, to perform *any* promise that forms the whole or *part* of a contract. See Contracts and Consideration, page 140.
Building Control	A service generally limited to periodic review of construction progress for certification to the lender of the progress of the work and the appropriate amount of funds to be released for payment to the contractor(s).
Budget	Budget may be used as a noun or as a verb. A budget is a projected cost for a defined scope of work. It may apply to a complete project, or to a specific scope of work.
C+FF	Cost Plus Fixed Fee (also CPFF): A contract for services which fixes the fee to be paid to the party performing the services defined by the contract, leaving the direct, indirect and reimbursable costs of the work to be performed under some other control, such as competitively bid lump-sum contracts, Time and Materials, etc. See chapter 19.
CPM	The Critical Path Method. See chapter 14 and A-2.
Certificates	All sorts of certificates may be issued on a construction project. Certificates of manufacturer's compliance; certificates of origin; certificates of completion; certificate of occupancy. The certificate is intended to assure the receiving party that some other party has checked and has confirmed what the certificate warrants. Parties offering services may also produce certificates of their training and competence.

Change Order	Also "Contract Modification." A written authorization by the owner that modifies the scope and/or conditions of work (but not always the time or the costs) from those defined in the original Contract. May apply to design, construction or any service contract.
Change Proposal	The quotation proposed for a potential change in scope to the current Contract terms and conditions.
Claim	Refers to a dispute formally filed in a court of law. It is often loosely used to describe a dispute prior to such formal filing.
Close (a verb)	The term is used to mean that all the dimensions that define a building or layout of a site come together, or "close" within acceptable limits for accuracy.
Compensable	Refers to asserted damages if two conditions are met: (1) the cause of the damages were beyond the control of the party asserting the damages; and (2) the damages are not excluded by the terms of the Contract (as for instance damages due to strikes, Acts of God, etc.). The opposite term is "Non-compensable."
Compression	Another term of art associated with construction claims. Its purpose is to suggest that by forcing too much work into too short a time (as result of delays, acceleration, ripple, etc.) inefficiencies have been imposed on the contractor's performance of the Work (the proverbial 10 pounds in a 5-pound bag).
Conceptual Drawings	See Schematic.
Conceptualize	In this book, the word is used to suggest an intensive effort to put into three and four dimensions (when time is considered) all of those activities and physical relationships that will go into developing a building project *before those activities are actually started*. The ability to do this requires many of the attributes described under Management (chapter 7) and in chapter 22. There is a direct correlation between how effectively a project can be *conceptualized* and anticipated in the Contract documents and the ultimate success of that project.
Concurrent	The modifier usually occurs in response to a construction claim when one party asserts that work activities on the project have been *impacted* by another party's breach of contract. If the party claimed against is able to demonstrate that the claiming party was responsible for impacts to the project that occurred over the same period—*i.e., concurrent impacts*—the claiming party's assertions may be reduced to that extent. The concept is usually associated with *delays* and *concurrent delays*.
Conformed	Construction Documents modified and updated to incorporate the changes in design or actual construction that may have accumulated up to an identifiable point in time. (Discussed in chapter 9.)
Consideration	The exchange of promises and/or performance within a Contract. See extended discussion in chapter 11, page 140.

Construction Documents	(CDs) The documents that describe *what* the Contract requires in the completed project. The CDs also define *how* the project is to be administered. The CDs are essentially the drawings and specifications incorporated into the Contract; but may include other essential documents such as the Instructions to Bidders, soils reports, etc.
Construction Management	(CM) The term has become so loosely interpreted and implemented that contemplation of what it means can only be defined by any respective owner after careful consideration of the description of management in chapter 7, and of chapters 16 and 17.
Constructive	A modifier that may be used with any number of conditions leading to an enforceable modification to a contract. *Examples:* "Constructive change order" refers to conditions which, under the terms of the contract, warrant a change in the contract scope and terms, whether or not the parties have explicitly agreed to the change. "Constructive acceleration" results from conditions whereby the party performing under the contract was required to increase its resources in order to complete within contractual limits because an extension to the contract duration was not provided, despite provable facts that conditions warranted such an extension.
Contract	All of the terms and conditions that are contemplated under an agreement for services. The specific identification of the documents that compose the Contract need to be spelled out in the Contract.
Contract Administration	The management of the performances required under a Contract. See chapter 7.
Contract Documents	See Contract.
Contract Modification	See Change Order.
Contractor	A party that commits to perform services under a Contract. Most owners limit their concept of the word to the general contractor for construction. Any party, however, performing under a Contract—architect, consultant, service company, etc.—is literally a contractor. If there is a need to be specific, the word should be preceded with a specific modifier.
Controlling work	The controlling work is that sequence of design or construction activities which can be shown at any time in the prosecution of either the base contract or extra work to be any work, the performance of which will determine the completion date of the Work under the contract. The controlling work can include administrative performances, such as the submittal or review of shop drawings; and/or the delivery of materials. The controlling work in a CPM schedule is referred to as the "critical path."
Cost Report	A report that accumulates into categories that have been pre-identified in a budget, the actual cost experience in each of those categories. A cost report is the means of implementing one of the important measurements of progress that needs to be made in the course of construction. The distinction between a cost report and an accounting report is that the first should serve the purpose of continuing measurement while something

A-1

can still be done about the *direction* of costs, while the latter is a statement of past costs about which there is little that can be accomplished except to argue about what *should* have been done.

Court	The venue in which trials are heard and judgments rendered. The word "court" is also often used as a synonym for the "judge."
Credentials and Experience	It is worth mentioning that certificates and credentials are often mistaken as assurances of competence. There can be a broad abyss between credentials and competence, between certificates, experience and reliable guarantees.
Cross Section	Also "section." A view of a building seen as though a knife blade had been passed vertically through the building at that point. (See page 123.)
Damages	A term used to describe or quantify the costs experienced as result of delay, inefficiency or other impact experienced by a party to a construction dispute. Quantification and proof of damages is fundamental to the presentation and/or defense of a claim.
Database	A system—manual or computerized—designed to receive and store information in such a way that it can be organized, selected, sorted and retrieved in different groupings. The use of a computer database is discussed on page 325.
Delay	The term usually describes extension of the duration required to perform work for which the basic scope and duration are defined by the Contract Documents. The assertion of delay may relate to a specifically limited scope of work, or to the Work of the entire Contract. See chapter 15.
Deposition	The testimony of a witness taken upon interrogatories, not in court, but written down and authenticated, and thereby admissible at trial or arbitration.
Design-Build	The contracts for these combined services take many forms. The process usually brings the general contractor and subcontractors directly into contract with the owner, and these parties retain design consultants in their respective disciplines. There is nothing fixed about this concept however, and the formats under which design and construction are undertaken concurrently are being continuously modified to meet different owner requirements. Design-build is to be distinguished from design-bid-build, which requires that design be completed, at least for a specific discipline, before construction bids are invited. See discussion in chapter 19.
Design Development	The design phase between Schematics and Working Drawings, in which the basic and controlling dimensions of the project and principal materials and equipment become fixed. The technical specifications (Divisions 2 through 16) should be (but frequently are not) substantially advanced in this phase.
Direct Costs	Costs expended directly into the construction of the Work in the construction documents. Broadly equivalent to "hard costs."

GLOSSARY

Discovery In a litigation the term is used to include any of the several devices that can be used to obtain facts and information about the case from an opposing party. These devices include (but are not necessarily limited to) interrogatories, depositions, production of documents, examinations of witnesses, or requests for admission of evidence.

Dispute Usually, in construction, an argument between parties to a Contract about the terms and conditions, or the performance under that Contract. Can refer to *sub*contracts as well as to design and construction contracts. In this context the word is usually applied to unresolved differences *up to* the filing of a formal claim.

Disruption Refers usually to a condition that has impacted the efficiency with which the contractor is able to perform the Work of the Contract. It is to be distinguished from, but is often confused with, Delay. Disruption may or not cause delay to the overall duration of a Contract. One way to test the assertion of *disruption* is to determine whether conditions beyond the control of the contractor required the contractor(s) to employ resources in ways not contemplated under their original work scopes and schedule. See chapter 15.

Downtime A period of time within the Contract duration over which no physical activity is being prosecuted. Downtime resulting in non-productive labor hours on a project, or "idle equipment" can be very damaging to a budget if one party can establish that another was the cause of the non-productive time.

duces tecum "Bring all the documents you've got." Often associated with a subpoena or demand for someone's deposition.

Elevation A surface view of a building which shows the external surface of the building (to be distinguished from "cross section").

Engineer [1] One of the designers working in conjunction with the architect to design the work to be constructed within a separate discipline. e.g. electrical engineer, structural engineer, mechanical engineer, etc.

Engineer [2] The lead consultant whose duty it is to design the overall elements of an engineering project, and to coordinate the services of the other design consultants whose design elements will also be part of the project.

Engineer [3] Sometimes used to describe the owner's representative for administration of the construction contract, usually on an engineered project such as underground utilities, highways, bridges, etc.

Entitlement A condition that attaches to a complaint once the assertions underlying that complaint are proven. See extended discussion after page 214.

Estimate Can refer to any of the estimates of quantities and labor prepared by the architect, the owner, a cost estimator or contractor at various stages of development, to project and/or confirm a budget.

375

Excusable	A condition which arises during the performance of a Contract that has impacted that performance and for which the cause or causes were beyond the control of the party impacted. See also "compensable." This is one of the conditions for compensability of an asserted impact. The opposite term is "non-excusable."
Extended Overhead	The overhead costs resulting from extension of the duration of construction. Typically, it includes the overhead of local- and main-office costs associated with the administration of the party whose contract has been extended. Because administrative overhead is not typically reported in relation to specific Contracts, a contractor assumes the responsibility for establishing a reasonable basis for allocation of overhead costs associated with delay on a particular project. Different formulas have been advanced, the most frequently asserted being the "Eichleay Formula," about which an owner should have a discussion with both his guide and attorney.
Extension	The word is used in estimating to identify the dollars that result by multiplying the quantities of labor, materials and equipment that are estimated to be employed in the performance of a work activity, by the unit costs associated those quantities. The aggregate of all cost extensions becomes the overall estimate for the Work.
Extra Work Order (EWO)	A system whereby a contractor may start to catalogue and identify potential extra work for which he may submit a request for additional compensation or time extension. (See pages 40, 103.)
FF&E	Furniture, Fixtures and Equipment.
FPO: Field Proceed Order	An authorization to a contractor to proceed with work in advance of the execution of a change order.
Fast-track	Any of several combinations of design and construction contracts which allow the construction to proceed while the continuing design of the project is still in process. See chapter 19 for some extended discussion.
Float (also "Slack")	When the sequences of tasks are calculated throughout a CPM schedule, a path or paths will be demonstrated as the longest continuous sequence(s) of work from start to completion of a project. This (these) will be identified as the "critical path(s)" of the project. Other sequences of tasks will necessarily have shorter cumulative durations in relation to the critical path(s). The difference between the duration of any shorter path and the critical path is called "float" (sometimes "slack").
Force Account	Similar to Time and Materials. The term is often incorporated into public-agency contracts, where it is usually accompanied with extensive requirements for measurement and documentation of the costs expended.
Front-Load	To *"front-load"* a project means to allocate payment amounts in the contractor's payment breakdown so that the contractor, if paid these amounts as scheduled, will be accumulating payment at a rate faster than the work represented by those payments is accumulating costs. See chapter 14.

General Conditions (1)	The General Conditions define the basic requirements that will govern the contractors' and others' performances under a construction contract.
General Conditions (2)	The term is also used by contractors to mean the costs associated with their administration of the contract. It will sometimes be incorporated into a construction contract as a percentage to be applied as markup to change orders.
General Contractor	The construction-contracting entity that is in primary Contract with the owner to perform the services of coordinating and managing all of the resources for producing what the construction documents require.
GMP	Guaranteed Maximum Price. The concept of a GMP is discussed in chapter 19. (See pages 76, 88, 89.)
Hard Costs	Usually refers to the costs of the labor, equipment and materials that are installed in, or required, to support the construction of a project. See also "soft costs."
Hearsay	Information not directly perceived by the person testifying, but instead learned through others. See discussion in chapter 13. See also "percipient," below.
Hierarchy of Documents	In order to avoid the potential for conflict among the various elements of the Contract Documents, the primacy of documents should be identified in order to provide a reasonable basis for resolving conflicts in the documents.
Impact	This word has become almost a term of art in construction litigation. Its use is different from its usual meaning in the non-construction world. It refers to the hurts, bruises, costs, delays, inefficiencies and almost any other complaint to which a contractor may refer in a construction claim. *Impact* can be either a noun or a verb in these contexts. (e.g., "it *impacted* the contractor's performance.") There can be "delay impacts," "cost impacts," "impacts to efficiency," "morale impacts" (to labor forces) and all manner of impacts about which the contractor may complain in the course of attributing causes and effects to his asserted inability to perform under the terms of the Contract. See chapter 15.
Indirect Costs	If the direct costs are those associated with the labor and materials that go into a project, the indirect costs are the administrative and overhead costs associated with managing, coordinating and documenting the direct costs. See chapter 15.
Inspection	In its specific sense, *inspection* is the effort to ensure that the design intent of the Contract is complied with. There are several levels of assignments within this context. The owner may have one or more inspectors who are to ensure that the owner will be delivered the quantity and quality for which he's paying. The local building agency may have an inspector to ensure that codes are being complied with. State agencies may require that the owner retain an inspector responsible to confirm that state regulations are being met. The owner needs to address the specific scope of work for the inspection work the owner pays for in order to

ensure its own requirements are met. Agency-directed inspectors will not protect the owner's concerns. See chapter 8.

Inspector	Note that inspectors from governing agencies have no duty to inspect to ensure the quality of the work in relation to the specifications, nor responsiveness to the owner's specifications. Their sole concern is that the work complies with applicable codes.
Interrogatories	A series of questions propounded in a prescribed format by one attorney to the attorney of the opposing party. The ostensible purpose is to make more efficient the gathering of information that could lead to resolving a claim. The practical use of the device by attorneys is to expand the time, the inconvenience and the costs of the opposing party to the extent that it is deemed to benefit the propounding attorney's strategy. As a realistic matter, interrogatories seldom produce information of any substance in construction-claims matters. See box on page 171.
Job-Walk	It means to *walk the job*. The term is subject to wide interpretation. To some, it means just eyeballing the immediately evident condition of a project. To others, it means to scrutinize the project to determine the standards of performance being met, or the adequacy of the work-in-place, or the efficiency of the logistics being employed for construction. It is one of the terms which, if it is to be relied upon for contractual enforcement, must be spelled out in detail.
Learning Curve	In most labor-intensive work, the efficiency of labor improves with time. This can be represented by a curve that reflects an increase of productivity over time. The curve is often referred to as the "learning curve."
Lien	A claim filed upon a property that, subject to proof of the conditions warranting the lien, entitles the filing party to encumber the property as security for the amounts and/or damages the claimant is owed.
Litigation/Court	A shorthand for describing the arena in which the issues related to a dispute are heard. Litigation is the broader term. (The) "Court" sometimes is a synonym for "the judge."
Log(s)	A log is a record of the comings, goings and processing of paperwork that passes between the contractor(s) and design consultants: submittals, RFIs, ASI, change proposals, etc. Accurately maintained on a regular basis, the logs are as important as daily reports, change orders and meeting notes. That is why the owner should insist that they be scrupulously maintained and forwarded on a regular basis into the owner's files.
Loss of Productivity	Performance of an identifiable scope of work at a level of efficiency less than planned. The proof for loss of productivity requires that the original rate of performance be proven before the relative loss, and the causes for that loss, can be established. See chapter 15.
Lump Sum	A fixed dollar amount proposed or agreed to, for the performance of a specifically defined scope of work. Can refer to the amount in a change order as well as in the original Contract. See chapter 19.

Measured Mile	The term is usually associated with a disruption claim (chapter 15). It refers to the base productivity against which a contractor will compare the asserted impacts on his efficiency or productivity.
Measurement	The evaluation of performance in relation to a previously defined standard for that performance. May refer to measurement of time, productivity, costs, or performance in relation to identifiable scopes and durations of work, and to standards applicable to respective parties' commitments ("professional standards," "standards of the industry," etc.) Different standards require different tools for measurement.
Mediation	An arena for resolving disputes whereby the parties mutually agree to have an independent party assist in resolving the dispute(s). Mediation is frequently undertaken by the parties only after a formal claim has been filed, but this is not a requirement. *BIT* recommends that mediation and negotiation be given serious consideration by the parties before undertaking legal action. See chapter 13.
Meeting Reports	The participants on the owner's project will necessarily meet frequently to review progress, resolve design issues, confirm payments, update and adjust schedules, attempt to resolve disputes. Their meeting will frequently raise issues that will require timely responses, and follow-up decisions and actions. Because any or several of the participants can be responsible for the responsiveness of these actions, the owner needs to have the records of these meetings maintained by someone who is objective and responsive to the owner.
Mitigation	It is the responsibility of the contractor, when he experiences an impact to his performance under a Contract, *to mitigate* the damages consequent to that impact. The law recognizes a doctrine of "mitigation of damages," which imposes on the injured party a duty to exercise reasonable diligence and ordinary care to minimize the damages after the injury has been inflicted. In construction, this means that a contractor cannot take advantage of an impact to his performance by further delaying the Work, or otherwise exaggerating the consequences of that impact.
Multiple Prime(s)	A system of Agreements executed between the owner and separate parties (therefore "multiple" or "segregated") for services. The presumption (not always realized) is that the aggregate of these agreements will perform the entire Work required to complete the project. Also "Segregated Prime." See chapter 19.
Notice of Completion	Notice filed, usually with the local agency that controls building codes, that the project has been completed.
Notice to Proceed	Notice issued by the owner which fixes the date on which the duration of the Work defined in the Contract will start.
Overhead	Administrative costs (other than field costs, which are considered Direct Costs) expended on the administration of Contract Work. See also Extended Overhead.

A-1

Owner	The party who/which pays the bills and owns the beneficial interest in the project being developed. As used in this book the word "owner" may shift back and forth between the party that has title to the project (an individual, a board acting for ownership, an agency acting on behalf of the public, etc.) to the individual who has primary responsibility for ensuring the success of a respective project. The context (hopefully) will make it clear which usage is in operation. See all chapters, and specifically chapter 21.
Owner's Representative (OR)	The person or firm retained to be the owner's alter ego for advice, evaluation and/or administration in relation to the design and construction contracts. The OR's authority may be limited to monitoring and recommending, or may be expanded at the discretion of the owner. The OR is not typically empowered to sign contracts or modifications. See chapter 19.
PSF	Pounds per square foot. A standard of measurement that relates typically to the loads imposed on horizontal or vertical elements of a project (soils, foundations, walls, roofs, etc.)
PSI	Pounds per square inch. A standard of measure that usually, in construction, represents pressure or strain.
Partnering	A term which, like CM, requires extended discussion. See chapter 16.
Percipient	One who perceives; i.e., someone who was present at and became directly aware of an event or condition. In construction-claims matters, the testimony of a *percipient witness* is limited to testifying about what he or she saw, heard or experienced of his own direct cognition. This is distinguishable in the law from *hearsay*.
Pre-Bid Conference	Once the owner issues invitations for contractor bids, there is usually a need to assemble the potential bidders to review conditions in the bid documents and to answer questions raised by the potential bidders. Sometimes the conference is combined with a visit to the jobsite. Chapter 7 and 9.
Pre-Construction Conference	After the construction contract is awarded to the contractor, and often before the contract is executed, there is a need to assemble the design consultants, contractor, and owner to discuss and confirm how the project will be administered, measured and documented. Chapters 7 and 9.
Prime Contractor	Usually the general contractor whose Agreement commits him/it to be the principal coordinator of all construction activities for the project. Also, in the case of Multiple Primes, each contractor that executes a Contact directly with the owner. See chapter 19.
Probable Construction Cost	The AIA design agreements commit the architect to provide the owner with the "probable construction cost" or "estimated construction cost" of the project at the end of each phase of design. See extended discussion in chapter 18.

GLOSSARY

Project Manager A person or firm with broad responsibilities for administering and directing development of the project: planning, design, funding, construction, etc. May be on the owner's staff, or may be an independent contractor or consultant. See also "Program Manager," chapter 17.

Quantum A word sometime used to mean the quantity of dollars that represents the damages experienced by a contractor on the project. The elements which, multiplied together, represent asserted damages. Those elements are typically time and unit costs.

RFI Request for Information (or RFC: Request for Clarification). (See pages 36, 52, 100.)

RFP Request for Proposal.

RME Responsible Managing Engineer

RMO Responsible Managing Officer

Record Documents Also: "As-Built Drawings." The drawings and specifications conformed at the end of the project to show the final locations, dimensions and specifications of materials and equipment as they were actually installed, or modified by change orders, where that differs from the original documents. See chapter 9.

Reimbursable Costs In almost all contracts for services these is a provision for the payment (with or without markup) of certain costs it can be anticipated that the performing party will expend in the course of its services, but which cannot, or have not, been estimated in advance. Reimbursable costs may include specifically limited expenses; or may, in some contracts, include the entire cost of the project beyond an agreed fee. Some Contracts are so artfully drawn that the performing party can readily ensure compensation for costs to which it should not be entitled, or to compensate itself for oversights in its base contract. Some contracts can also be written so that "reimbursable costs" include *everything* that goes into a project. It is an area that needs very close definition and measurement in every Contract, and may warrant open-book accounting under some contract formulas.

Retention Money held back by the owner from payments to the contractor. The limits and terms for retention need to be absolutely clear in the construction contract or both parties may be confronted with the basis for a dispute. Many Contracts are far from clear about the amounts that will be retained in the course of construction.

Ripple This has become another construction-claim term of art that accompanies *impact*. It is typically used to include and/or describe the consequential impacts to the performance of the Work that result from an event earlier in the project. As for example, the delay to early concrete work may have a *ripple* effect on the ability of other trades to perform their work. See chapter 15.

A-1

Schedule (1) A schedule in construction can be many things: (1) a "hardware schedule," which is a listing of the hardware components on a project; (2) a "window" or "door" schedule; (3) a "schedule of values," which is a listing of unit costs or other values associated with a project's budget or costs; or (4) an "equipment schedule," etc. As these phrases suggest, the work *schedule* needs either a qualifier or a clear context in order to define what kind is being discussed.

Schedule (2) A "project schedule" or "time schedule" (CPM, bar chart, etc.) is the kind of schedule discussed in chapter 14.

Scheduler A singularly problematical oxymoron. A "scheduler," as the construction industry uses the term, means someone qualified to develop a computerized CPM schedule. The qualification "scheduler" does not necessarily guarantee that the schedule itself is either competent or representative of a competent analysis of the Work. The question raised in *BIT* is whether someone who is not fully conversant with the processes being scheduled is qualified to organize the information necessary to plan, perform, control and report those processes. See chapters 14 and 17.

Schematic Also: "Conceptual." The designs provided in the first of three phases of design, after Programming. Schematic drawings typically describe the general concepts, ideas, and overall organization of components to be incorporated into the final design of the project. Schematic design may include renderings. Schematic design should—but often does not—include outline specifications. See chapters 7, 8 and 9.

Section See Cross Section.

Segregated Prime See Multiple Primes. See also chapter 19.

Soft Costs Those costs, essential to the development of a project, that are *not* directly related to the labor, equipment and materials that go into the project. Soft costs cover a very wide range of costs and charges on a project, including (but certainly not limited to) design fees, permit fees, financing costs, administrative costs, consulting costs, inspection, pre-construction investigations, the cost of preparing as-built drawings, etc. The failure to take all soft costs adequately into consideration when preparing a budget can lead to some very big surprises for an owner.

Special Conditions Or "Supplementary" Conditions; or something similar. The General Conditions (GCs) define how the organization (the owner's organization, or the AIA, or some equivalent generator of the basic administrative specifications) typically does construction business (see GCs). Every project is different, has its unique dates and durations, etc. It is usually necessary to expand the requirements of the GCs to define the conditions appropriate to a particular project. This is done with the "specials" or "supplementals." See chapter 18.

Special Masters (See pages 166, 175.)

GLOSSARY

Specifications The written instructions which typically accompany the construction drawings on a project and direct the contractor(s) in narrative form how the work depicted in those drawings is to be executed. There are usually two parts to the specifications: Divisions 0 and 1 (the boilerplate and specials) and Divisions 2 through 16, the technical specifications. See chapter 9.

Spreadsheet A system for organizing essentially numerical quantities so they can be calculated, accumulated and analyzed. A computerized spreadsheet can be a valuable tool for analysis of budgets, costs and other quantities that accumulate on a construction project.

Stacking As in "stacking of trades," a condition often associated with assertions of *acceleration* or *compression*, used to assert an argument for loss of efficiency as result of these impacts.

Stop Notice (1) A formal filing of a notice to a source of funding by one party to a Contract that the party responsible for underwriting that funding (say, the owner in relation to a bank as funding source) had not met the requirements of the Contract, and that the filing party is thereby demanding that funding on that project be held up.

Stop Notice (2) A notice to the contractor to stop the work of the project issued by someone (the owner, the CM, the inspector, etc.) with authority to do so. The use of a Stop Notice must be exercised with careful discretion because of its potential for initiating a dispute about delays to the project.

Stranger Someone you would not allow to control your checking account. Used in this context, "stranger" does not necessarily exclude acquaintances. It is used in this book essentially to emphasize a level of trust both in someone's integrity and in his or her capacity to represent the owner objectively and competently in the matters required for delivery of a successful project.

Subcontractor A party not contracting directly with the owner, but whose services for design, construction or other work will become part of the project through the terms of an agreement between the owner and a Prime Contractor (general contractor, architect, etc.)

Submittal Anything the contractor is required to send to the architect for his review and acceptance in order for the general contractor to perform the Work under the Contract. May include schedules, shop drawings, field drawings, change proposals, payment requests, etc. (See pages 39, 52, 99.)

Superior Knowledge Information about the conditions surrounding the project (e.g., soil, prior construction, hidden obstructions, practical restraints, etc.) about which the contractor needs to be informed in order to perform under the Contract. Knowledge that is available to the owner but not readily apparent to the contractor unless the owner provides it or identifies where it can be found.

TQM Total Quality Management: a panacea discussed in chapter 16.

Time & Materials (T&M)	An agreement whereby a contractor will perform work while maintaining records of the actual labor and materials expended and will invoice accordingly. A form of contract that is hazardous unless specific scopes of work, and well-defined limits are established in advance. Even then it is a difficult and costly form of contract to administer because it requires close and careful monitoring and regular (daily) sign-off of confirmed expenditures to ensure accuracy and honesty. See chapter 19.
Tort	A civil wrong or injury *other* than a breach of contract.
Turnover	For Julia Child this has one meaning. In construction, it usually refers to the rate at which labor forces leave the project and are replaced with new people within the labor force of a particular trade or subcontractor. One reason *turnover* has a major significance in construction is that it has a direct relationship to *Learning Curve*. As such, it can contribute to accumulating losses of productivity.
Value Engineering (VE)	Described at some length in chapter 16.
Vendor	Also: Supplier. The folks who sell the materials and equipment to the contractor that he will install and that the owner will live with for the next 40 years.
Wasting Policy	An insurance policy or surety-bond policy may contain a clause that provides that costs expended by the surety in defending legal actions will be charged against the coverage value of the policy. A party who prevails in an action to recover the policy amount may find that it has been substantially reduced before the judgment is rendered. See chapter 12.
Weather	In construction, the term usually refers to rain, but could refer to any condition that demonstrably impacts the construction process. The word is often combined with inadequate modifiers like "normal," "unusual," etc. If it is to have any meaning with regard to construction impacts, it must be qualified in terms of whether it has *actually* impacted the construction work, and what "normal" means in the context of a particular project, in a particular location at a particular time. It can be equally important to restrict, or describe, possible residual effects of weather on the construction work.
Work	Work has the usual meaning in construction—product resulting from effort. In a construction contract, when it is spelled with a capital "W," it refers to the entire body of effort and services required to deliver a completed project. Sometimes used to mean the project itself: *"The Work."*
Work Item	See Activity
Working Drawings	The final construction drawings from which the project is to be built. The term is sometimes used to mean all of the design documents, including specifications, incorporated into the Construction Documents.

A-2
CPM—THE CRITICAL-PATH METHOD

Background

A man named Henry Gantt is credited with inventing in the late 19th century a simple and useful tool for scheduling work. His tool is called the Gantt chart, or bar chart. We first looked at bar charts in this book in chapter 14.

The bar chart was used extensively by both construction and industry into the early 1950s.

Simple as the bar chart is, it has valuable applications on many types of projects. It is easy to read. It conveys information quickly. A bar chart showing scheduled work can be overlaid with a bar chart that shows how that work was actually performed, in order to compare performances.

Few in management, and few laborers in the field are trained to understand critical-path schedules. Bar charts provide a useful format for short-term scheduling and for communicating information to the field—provided they're maintained as part of a larger, total-project schedule.

Even large, complex schedules, kept current and accurate through updating of a competently detailed CPM, can be plotted in bar chart formats for ready communication to the field and for evaluation by management.

A bar chart is a useful starting point for schedule development, permitting progressive addition of information and adaptation to CPM as detailed information accumulates.

The *intelligent* marriage of CPM, computers and bar charts can present useful information on projects that have hundreds, or thousands, of work activities. As the information in a CPM analysis is prepared for input to a computer, codes can be added to identify the responsibilities and locations of specific work.

When this information is input to a computer, the computer will correlate the logic of work sequences, add overall durations, assign dates and calculate "float" (see below). It can be instructed to select and sort information into bite-sized groups that can then be printed (plotted) in separate bar charts, so that separate charts with selective information can be distributed to respective trades, or crews or responsible parties, or used to report work in specific locations.

The Limitations of the Bar Chart

The bar chart, however, has severe limitations as a basic scheduling tool once the number of work assignments in a project exceeds 40 to 50 work activities.

The first limitation that's apparent in a bar chart is that the relationships between the respective work activities are not shown, even though it's evident that some work starts at the midpoints of other work. The relationships that control the logic of these intermediate dates are not explicitly shown. As a result, there is insufficient information with which to assess the resources required, or to identify the controlling work of a project.

P.E.R.T.

In 1957-58, a new invention was introduced to facilitate this kind of analysis. It was called P.E.R.T. (Project Evaluation Resources Technique). P.E.R.T. introduced a way to make explicit the relationships implicit in the bar chart.

Once these relationships became explicit, P.E.R.T. made it possible to break work activities down into more, smaller, shorter, resource-dictated pieces, so that the "path(s)" of activities that "controlled" the overall project could be calculated.

Once the controlling activities along this path (or paths) were identified, their cumulative durations could be added to define the overall duration of a project. Then, the relative durations of other paths could be calculated.

The differences in the aggregate durations of other paths in relation to the overall controlling path were identified as "float" or "slack." This represented the amount of time (usually in days) by which these paths could be extended or otherwise changed without affecting the overall project duration.

This series of calculations, in turn, provided a way to analyze the costs and the value of adjusting the resources on some of the activities. This provided a very effective way to measure which changes in logic, durations or resources would result in an optimum schedule.

CPM—THE CRITICAL-PATH METHOD

P.E.R.T.'s First Project—A 50% Reduction in Time

The value of this potential to *analyze* the specific relationships, durations and costs associated with respective work items—and to determine which of them were the *controlling items*—was immediately demonstrated on the first project to which P.E.R.T. was applied. A four-year military project was completed in two years.

The second project on which P.E.R.T. was applied was a commercial one. The lining of a catalytic-cracking tower at a chemical-processing plant had never before been changed over in less than 21 hours. The first time P.E.R.T. was applied to analyze this process, the changeover was completed in 12-1/2 hours. The second time in 11 hours. And the third time in 9 hours—a reduction in time of 57%.

Enter the Critical-Path Method

The techniques basic to P.E.R.T. were quickly picked up by the construction industry. By 1961, it was being used on diverse construction projects. In construction use, it was renamed CPM: *The Critical-Path Method.*

"Critical Path" became the shorthand for the "controlling work path." It denotes the fact that, for every project *that is adequately detailed,* there is a discrete number of work items (usually on the order of 5% to 15% of the total number of work items) that *control* the overall duration of the project. They form the *critical path,* or *paths,* of work items that must be performed as scheduled if the project is to meet its projected completion date. This path, if lengthened, delays the completion. It is also the path which must be shortened if the project is to be delivered earlier.

The idea of *path* comes from the way the CPM chart is drawn. Work items are connected end-to-end in the order, or sequences, in which they must be completed in order for succeeding work to be performed. The result is a *network diagram.* The sequences that are continuous between any two selected points are called *paths.*

With these basic ideas about what CPM can do, it's time to review what it is, and how to use it.

A-2

HOW DOES CPM WORK?

Defining "Activities" and Their Relationships

Figure A2.1

B is a CPM activity drawn in an "arrow" or "line" diagram format.

This part of a network diagram, or analysis, says:

- **B** can start <u>only</u> after **A** is completed.
- **H** can start once **B** is completed.

The points (circles) between the activities are "nodes." Sometimes they're called "events" because they represent a point in time when something has been completed, or can start.

d is the duration of **B,** usually in workdays in construction.

If we make this partial diagram a little more complicated and put in workday durations for each activity, we can start to calculate the cumulative durations for a small project.

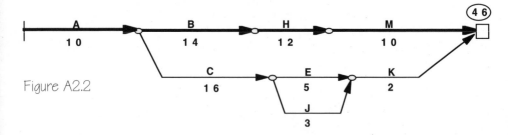

Figure A2.2

If we add all the lines **A+B+H+M,** in Figure A2.2, we get a total of 46 workdays.

B+H+M taken alone equals 36 workdays.

C+E+K equals 23, so this partial "path" is 13 workdays shorter than **B+H+M.** Since there are no other activities in this brief diagram (except **J,** which we'll get to), it's immediately evident that **A+B+H+M** represents the controlling, or "critical" work on this small project. The "critical path" is 46 workdays long.

388

The path **C+E+K**, a total of 23 workdays, has 13 days "float" measured in relation to the parallel work on the critical path (36 less 23). Activity **A** is common to both paths, but is part of the critical path, so it has no float.

E and **J** are done concurrently, but **J** requires 2 days less than **E**, so **J** has 2 days more float than **E**.

Float Is Not an Answer

It's a question—or it presents several options:

- If we can reduce any activity durations along the path **A+B+H+M,** we can shorten the project duration (up to 13 days) without even considering **C, E, J** or **K.**

- If the critical-work durations are not reduced, we might be able to reduce our resources (costs) and stretch out the durations of **C, E, J** and **K** up to an aggregate of 13 days.

- Or we could delay the start and/or finishes of **C, E, J** or **K** (one or more of them) up to a total delay of 13 days without delaying the overall project.

The Power of CPM

It is this kind of analysis that can make CPM such a potentially powerful tool.

The power of this analysis, however, depends upon a basic premise:

The CPM analysis must be developed at a level of detail that makes it possible to evaluate the adjustment of resources and durations for specific responsibilities, trades, or crews.

Two Small Tools
(Within the Larger One We Call CPM)

Few projects are as simple as the ones above. Most have more complex relationships that will require two more subtools (even before we turn to a computer to assist with calculations).

These are *dummies (or dummy restraints)* and *lead times.* The best way to define them is to show how they're used.

Dummy

As we continue to analyze work represented in the diagram, it becomes evident that **H** also depends on the completion of **C**. A simple way to show this without redrawing the diagram is to run a *dummy* between the node at the end of **C** to the node at the start of **H**. See Figure A2.3 below.

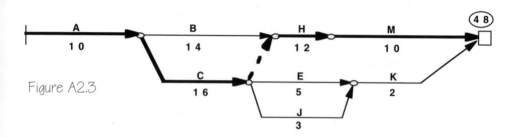

Figure A2.3

The dummy transfers the cumulative duration (10+16=26) at the end of **C** up to the node at the start of **H** *without adding any time.* We draw it as a dashed line to distinguish the *dummy* from activities that have durations.

By putting in this restraint on **H,** we have changed the overall duration of the project. We now have two cumulative durations to the starting point of **H**: A+B = 24; and A+C = 26. **H** now can't start until day 27, while before it could start on day 25. The project is two days longer.

The critical path now runs through **A-C-H-M = 48**. We will have to recalculate the float on all other paths. We'll get to that in a couple of pages.

For now, consider this technique: activities can be drawn in the "sequencing" that most clearly shows related flow of work activities. If it becomes necessary to show other relationships between separate paths, the dummy can be used to demonstrate those relationships.

When you discuss CPM with others you will hear about "predecessor activities" and "successor activities." In other systems, these are shown in different, and sometimes less-than-clear ways (e.g., **C** would have an FS—"finish to start"— relationship to H). These notations (FS, SS, SF, etc.) are not necessary in an arrow-diagram format because they're explicitly shown by the lines that run into and out of nodes. You can forget them, unless you have a need (which I've argued against for 25 years) to learn any of these less-than-explicit systems. In an arrow diagram, all you need to do is to think in terms of what lines flow into, and out of, specific nodes or endpoints of activities.

Lead Time

Now, let's change our diagram one more time to learn about *lead times*. Then we can start to pick up on calculating more complicated networks.

First, erase the dummy from the end of **C** to the start of **H**. We decided that relationship wasn't true after all.

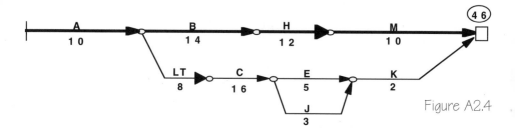

Figure A2.4

What we realized instead was that **C** can't start until at least 8 days of **B** have been accomplished. We don't want to split **B** into *Start B=8, and Complete B = 6*. That would break **B** into two parts that would be unintelligible to the people who are performing the work of **B**. See Figure A2.4.

So, instead, we run an activity from the start of **B** to the start of **C** and identify it as a *lead time* by putting an LT in the work description. We assign it a duration of 8 workdays. Now our diagram says *activity C cannot start until the first 8 days of B have been accomplished*.

We have changed the durations of the path **A-C-E-K** by adding a LT of 8 days, so this path now adds up to **10+8+16+5+2 = 41**. It is still shorter than the critical path **A-B-H-M = 46**, but we have changed the floats for all but the critical path (on which float = 0) and will need to recalculate them.

A-2

Shorthand Identifiers

We might, as the CPM network analysis becomes more detailed, want the assistance of a computer to add and subtract durations and to calculate float. Once we have learned how to count through the durations manually—that is, learned how to calculate dates and float through the end of a sharpened pencil—we might want to turn those calculations over to a computer scheduling program.

The computer will want a shorthand code to identify the work activities, rather than ingest the full descriptions of all work activities while it's working out dates and float. We can provide each activity with a shorthand identifier, and the computer can then relate these descriptions with the respective identifiers *after* it completes its calculations. We make the nodes larger to accept identifying numbers. Diagram A2.4, prepared for computer input, would look like this (Figure A2.5):

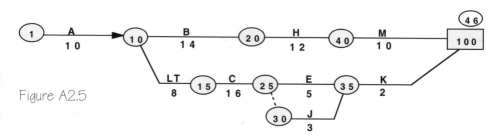

Figure A2.5

The computer identifier for **B** is 10-20, and for **H**, 20-40. Long ago someone used Greek symbols—"I" and "J"—to name these identifiers: "10" is the "I" number and "20" is the "J" number for **B**.

Since "20" is also the "I" number for **H**, this system of "I-J" identifiers automatically translates the logic of the relationships we've drawn to the computer. *Every J number has to have an I number leaving it; and every I number needs to have a J number coming into it* (excepting for the first and last activities in the network). Or, **Every J has an I, and every I has a J.**

As we numbered the **I-J** nodes for computer processing, note that we added a new node at the start of **J.** We ran a dummy from the node at the end of **C** to this new node, which transfers the same cumulative duration to the start of **J** that it had before.

392

This new dummy/node combination was inserted solely to provide **J** with a unique identifier: 30-35, so the computer would distinguish it from **E (25-35).** The rule for numbering a diagram for computer processing is:

Each activity must have a unique numerical identifier.

Note also that, since we have taken care to draw the diagram so everything flows to the right, we have been able to eliminate the arrowheads. Our "arrow" diagram is now a "line" diagram.

It takes more words to explain how the logic of CPM is developed than it does to draw it. This is the beauty of the arrow- or line-diagram system. If you have drawn the logic correctly to represent the relationships you have in mind, that's all it takes to use the computer to calculate the results. The computer knows all it needs to know by connecting *I*s and *J*s and adding up durations.

No hidden logic. No hidden notations. No games—because if it's not specifically shown in the logic-network diagram, it doesn't figure into the schedule. *You can't play the game of hiding restraints in the computer to make 83% of the work activities "critical."*

A Special Note About Systems and Abuses

Twenty-five years ago, when computers were clunkier, and there was a premium on computer memory, a computer jockey invented the "precedence" system for CPM.

Its very temporary advantage (about five to seven years before computer capacity caught up) was that by identifying activities with one number only, and substituting symbols for dummies and lead times, the memory load on the computer could be reduced. Fewer numbers, fewer activities meant (temporarily) that computers could calculate the network more quickly.

The precedence system produced diagrams with "time in a box" instead of "time on a line." The box had a single number; and the relationships (our lead time in Figure A2.5, for example) would be put into the computer as **B-SS8-C (C** follows **B** after a start to start offset of 8 days).

Only a computer could love annotations like this, instead of the explicitly drawn lead time in our diagram. In fact, because they can be hidden in the computer processing, their influence on a schedule can also be easily hidden from the receiver of that schedule.

For the benefit of efficient computer processing, the "precedence" system for developing a schedule diagram generated two abuses about which computer types were—and continue to be—unconcerned:

- It took away the natural way in which we all see time: as flowing from left to right, like a river, or across a calendar. It put time into a box. Lines were run between boxes, with often illegible notations such as "SF" or "FS."

- It made it possible for the unscrupulous (you count 'em) to hide symbols inside the data fed to a computer to create schedules that have (as the story on page 252 demonstrates) 83% of their activities critical, despite the fact there may be no supporting logic in the actual relationships of those activities.

The whole argument for the precedence system went away before the end of the '70s when computers stopped clunking. Minicomputers could handle almost any size of project. With PCs, you can do on the kitchen table what required a kitchen-sized room in 1965.

But the investment in software had been made and we continue to be victim to the computer industry, and the unscrupulous who take advantage of the ability it provides to hide stuff.

Now, many "precedence" diagrams are drawn looking like arrow diagrams, but their users and processors continue to employ the non-explicit nomenclatures ES, SS, SF, etc., etc., and to hide them in computer algorithms. Reason? Because the purveyors of these software systems only teach the "precedence" system, and because they have discouraged the use of their programs for an arrow/line diagram (which is still possible if you can get anyone to show you how).

The mouse and the lack of competent instructors in the arrow-diagram system has contributed to this, as we've already discussed in chapter 17.

The Significant Advantages
of the Arrow Diagram

- It looks the way we think about time: it flows.

- It requires a more rigorous thought pattern. Effectively developed, all relationships are drawn from left to right, with no lines running backwards. That requires a certain clear logic to keep things in the diagram in the same reasonable pattern that can be expected out on a project.

- It can be non-scalar. The actual length of an activity in a CPM logic diagram can be the same for a 3-day activity as for an 18-day activity. The computer will add the durations as numbers without regard to how long each line is drawn. The computer can then put it all back into a time-scaled presentation (bar chart) to communicate the information to the non-CPM literate.

- Complex projects can be developed on manageable sheets. Because the diagrams can be non-scalar, and separate sheets can be connected readily with explicit dummies or lead-time activities, it becomes possible to develop large schedules in efficiently sized drawing sets (similar to design drawings) and to follow the logic between separate sheets (which might represent separate buildings, or floors, or processes). This continuity is extremely difficult to follow with *time in a box* and non-explicit (or hidden) annotations like SF, FF, SS, etc.

- Since the logical relationships are maintained in a readily erased-and-redrawn non-scalar diagram, the schedule can be quickly updated.

- The bar chart the computer produces (see Figure A-2.6 on the next page for an example of one format) does not need to have any confusing lines or connections between activities. It can be returned to the easily read bar chart format, as long as at least one competent person can provide assurances that the logic of the diagram has been maintained through the computer input. (There are simple ways to confirm this.)

COMPUTERIZED BAR CHART DERRIVED FROM CPM ANALYSIS

WORKING DAY / DATE

	C			DUR	
I	P	J	ACTIVITY DESCRIPTION		
1000 *	1090		CLEAR & DEMO	12	cccccccccccc | | | | ► |
1010	1020		SHOOT GRADES	5	xxxxx- | | | | |
1025 *	1040		GRADE, SCARIFY & COMPACT	10	Icccccccccc | BAR CHART CONTINUES
1030	1040		STOCKPILE SOIL	5	| xxxxx- |
1050 *	1060		ELEC PRIMARY COND & VAULT	7	| | ccccccc | | | |
1090 *	1100		MAKE BLDG PADS	5	| | ccccc | | | |
1100 *	1200		EXCAV INTER FOOTINGS	7	| | | ccccccc | | |
1110	1130		UNDRGRD PLBG & BKFLL	10	| | | xxxxxxxxx-- | | |
1060	1120		EXCAV STORM LINES	6	| | | xxxxxx------ | | |
1070 *	1070		CITY PULL PRIMARY	2	| | | cc | | |
1210 *	1080		CITY REMOVE POLES	2	| | | Icc | | |
1160	1220		REBAR INTER FOOTINGS	3	| | | Ixxx--- | | |
1230	1170		LAY STORM PIPE	10	| | | Ixxxxxxxxxx------ | |
1250 *	1240		SET COL ANCHOR BOLTS	6	| | | cccccc | | |
1130	1260		POUR INTER TOOTINGS	3	| | | xxx-I | |
1140	1190		EXTER PLBG, LIFT STATIONS, BKFLL	15	| | | Ixxxxxxxxxxxxx----- |
1260	1150		BORE STREET & INTL STORM LATERAL	3	| | | xxx-------------- |
1170 *	1270		PLACE DRAIN ROCK	5	| | | ccccc | |
1300	1180		CONST CB'S & BKFLL	12	| | | Ixxxxxxxxxxx------ |
1150	1310		EXCAV & LAY FIRE LINES	14	| | | Ixxxxxxxxxxxxx------
1270	1180		CONST STREET STORM	4	| | | Ixxxx-------------- |
1270 *	1280		VISQUEEN & SAND	10	| | | | cccccccccc |
1280	1290		PLACE REINF MESH	10	| | | | xxxxxxxxxx---- |
1310 *	1340		CONST JOINTS & BLOCKOUTS	15	| | | | | cccccc
1350	1320		BKFLL FIRE LINES	9	| | | | | xxxxxx
1420	1360		POUR S.O.G.	10	| C = 1 work day critical item | | xx
1440	1450		FORM P/C PANELS	15	| X = 1 work day | |
1430	1450		SET DOOR FRAMES	15	| — = 1 day float | |
1320	1450		SET LEDGERS	14	| | |
1460	1330		INSTL SITE HYDRANTS	1	| | | | | |

	MAY	JUN	JUL	AUG	SEP OCT
PROJECTED PERIOD TOTALS	|	|	|	|	|
PROJECTED CUMMULATIVE TOTALS	|	|	|	|	|
ACTUAL PERIOD TOTALS	|	|	|	|	|
ACTUAL CUMMULATIVE TOTALS	|	|	|	|	|

COMPUTERIZED BAR CHART DERRIVED FROM CPM ANALYSIS

SORTED BY TRADE: EXCAVATION & UTILITIES

WORKING DAY / DATE

		C			DUR	
TRADE	I	P	J	ACTIVITY DESCRIPTION		
1	1000 *	1090		CLEAR & DEMO	12	cccccccccccc | | | ► |
1	1010	1020		SHOOT GRADES	5	xxxxx- | | | |
1	1025 *	1040		GRADE, SCARIFY & COMPACT	10	Icccccccccc | BAR CHART CONTINUES
1	1030	1040		STOCKPILE SOIL	5	| xxxxx- |
1	1100 *	1200		EXCAV INTER FOOTINGS	7	| | ccccccc | | |
1	1110	1130		UNDRGRD PLBG & BKFLL	10	| | xxxxxxxxxx-- | | |
1	1170 *	1270		PLACE DRAIN ROCK	5	| | | ccccc | |
2	1050 *	1060		ELEC PRIMARY COND & VAULT	7	| | ccccccc | | |
2	1090 *	1100		MAKE BLDG PADS	5	| | ccccc | | |
2	1070 *	1070		CITY PULL PRIMARY	2	| | cc | | |
2	1210 *	1080		CITY REMOVE POLES	2	| | Icc | | |

Figure A2.6

Calculating Durations, Dates and Float

We have one more small tool which we'll use while we're manually calculating durations and float.

I call this tool a "balloon." It's an ellipse that we'll put (temporarily) at the start and finish of each activity. The balloon looks like this:

late start or
late finish

early start or
early finish

Figure A2.7

When the balloon is located at the start of an activity (just after the node), it will contain that activity's start day, measured from day 1, the start of the project. When at the end of an activity, it will contain the accumulated finish days to that point.

The first thing to note is that the start day on each respective activity will be *one day after the latest of the finish days* that come into the other side of that node.

Or, to say it another way:

> **The start day of any activity is one day after the longest of the cumulative durations of all the activities that control the start of that activity.**

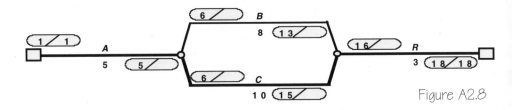

Figure A2.8

In our small project above (Figure A2.8), we have added up the cumulative durations along the respective paths and have shown the early finish and early start of each activity.

Counting along the two possible paths, it's evident that the critical duration goes through **A-C-R**.

Calculating the Late Days: The *Backward Pass*

We've been able to put in the latest start day of **A** and the latest finish day of **R**, because it's readily apparent that these must be the same as their early days because they're both on the critical path (neither has any float).

What we don't know yet are the latest start and finish days of the other activities.

We calculate these by performing a *backward pass* through the network. You can follow the development of this sequence in Fig. A2.9.

In making this *backward pass* there is one absolute rule that applies:

The *latest finish* day for any activity is the _earliest_ of all of the late finish durations that result from all backward passes to that point.

- Subtract **R**'s duration from its late finish (18). The result, 15, is the latest of the days that can *precede* the start of **R**. Since there are no other possibilities, 15 goes into the latest finish for *both* **B** and **C**.

- Subtract the duration of **B** from its late finish (15 - 8 = 7) and put this number *temporarily* above the balloon at the end of **A**. (It's temporary until we perform the same calculation for **C**.)

- Subtract the duration of **C** from its late finish (15 - 10 = 5) and put this number, also *temporarily,* below the balloon at the end of **A**.

- Since there are no other paths to calculate, we now know that the latest finish of **A** is 5, because of the rule above: 5 is the *earlier* of the *two possible late finishes of* **A**. Now, we can put 5 permanently into the right half of the balloon at the end of **A**.

If this sounds confusing, think of it this way: *If the date by which A completes is any later than day 5, it will delay the start of C.* That will extend the critical path.

At this point, we have three more balloons to complete: the *late-start days* of **B, C** and **R**.

First, we know that the float at the end of an activity must be the same as the float at the beginning of that activity. So we can use the float we've already calculated in the *finish balloons* to calculate the late start days of those activities.

- For **A,** the float is 0. If we hadn't already put the 1 in the right side of **A**'s start balloon, we would use the float (zero) to add it now.

- For **B** the float is 2. Therefore, its late-start day is 2 days later than its early start day, 6 + 2 = day 8.

- For **C** the float is 0, so its late start day is the same as its early start day, day 6.

- And we know **R**'s float is also zero, so its late-start day is the same as its early start day.

The completed diagram is shown ins Figure A2.9.

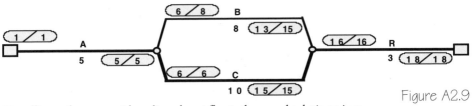

Figure A2.9

Recalling what we said earlier about float, these calculations just raised at least two questions:

1. Can we shorten **C** by two days? If so, we can shorten the overall project without looking at any other work items.

2. Can we stretch out **B** by two days and save money? If so, we won't affect the overall duration.

Making Sure We've Thought It Through

Let's add a few activities to our project. Here's a new diagram—Figure A2.10.

It has durations, but only blank balloons. The original four activities are in the network, with some new ones added. If you want to have a go at it, try filling in the balloons before turning to the next page.

Figure A2.10

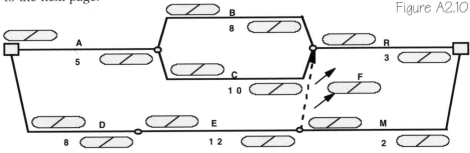

A-2

Here is the same diagram as on the last page with the balloons filled in.

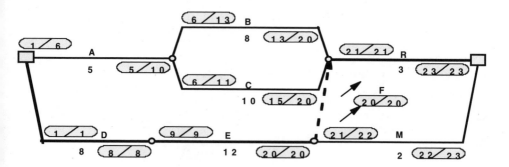

Figure A2.11

The critical path flows through part of the new work that was added to our original network — **D, E, F, R** — (note that a critical path can flow through a dummy). **B** has 7 days of float; **C** has 5 days of float.

But why use the dummy? Look at Figure A2.12.

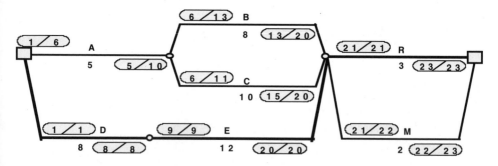

Figure A2.12

Why not run the end of **E** directly into the start of **R**, and then run M back out of the same node? Here's how the diagram would look:

The diagram analysis still looks OK. The start days, finish days and float are all the same.

400

But—what if, after we've changed the diagram logic, the durations of **C** and **E** change?

Look at what this does, in Figure A2.13. After we made this drawing change, thinking it did not change the effect of the scheduled relationships, the duration of **C** changes to 13 days, and **E** to 8 days.

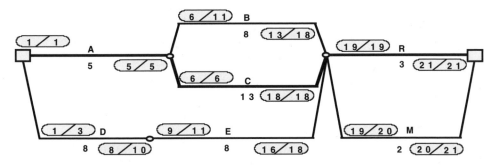

Figure A2.13

Now, although **E** finishes on day 16, and **M** should therefore be able to start on the next day, (as it did in A2.11), on day 17, it is now not able to start until day 19, not because of any true change in logic, but *only because of the way in which we drew the diagram.*

In Figure A2.13 **M** has only 1-day float.

So let's not run **E** into **R** directly, but return the diagram to its original logic (as in A2.11), so that **M** is <u>not</u> held back by either **B** or **C**, and see what the new durations should have produced:

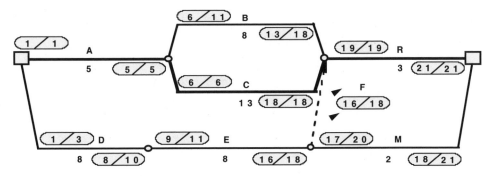

Figure A2.14

401

Now, while **E** still has only 2 days of float, because its late finish day must be 18 (in accordance with the rule above about backward passes), **M** can start on day 17. This accurately drawn logic shows also that **M** has 3 day's float.

> *It's important that the logical relationships of activities are developed, at least initially, independently of the durations that are currently assigned. Then, if durations are changed, either in the process of rethinking the project, or as result of progress or changes, the diagram will be as accurate as possible in relation to the sequences required on the project.*

Some Considerations That Follow From This Basic Introduction

1. You won't be required to perform these calculations on most CPM schedules. They can be done with any competent scheduling software system. I recommend that you find one that permits the use of the I-J numbering system before you even consider the precedence system. You can always make a shift later, but you will not have a background with which to assess rigorously the logic of the schedules with which you're presented if you start there.

2. In any competent scheduling software, you will have the facility to add coding for responsibility, trade, location, etc., and select and sort the output of the data accordingly.

3. Some scheduling software has tried to make its scheduling packages "user friendly" by asking only for the descriptions of work items, and then asking that you show how they're related (before, after, predecessor, successor, etc.).

    Here's an irony for you: once we had to cut the identifier down to one number to save memory. Now, "user friendly" systems use whole phrases to distinguish activities.

    Stay away from these programs. Construction is a numbers business. Numbers are more efficient, flexible and will provide you with better control over logic and revisions. If you're going to take any interest at all in benefiting your project, you'll have to be comfortable with numbers (budgets, reports, changes, etc.) anyway, so this is the time to commit yourself to a number-controlled software system for CPM.

4. A competent CPM software system will allow you to assign codes to the dummies and lead times in your schedule, and *no-opt* them in your charts and printouts. This means that while they will continue to have the proper influence in the calculation of dates, they need not be shown in (and thereby clutter up) your date listings and charts.

5. The translation of workdays to dates will be automatic. You will set up a calendar that spans the duration of your project, and will identify in it certain non-working days (holidays, etc.). Once the computer calculates the days to each respective start and finish day, it will assign and print the correlative dates.

6. The software will also allow you to assign resources, dollars, crew names or sizes and a considerable amount of other information to each activity if you have a need to, which will in turn provide you with automated cashflow charts, labor charts, etc.

7. The manual calculation system you have started to learn here will provide you and your guide with a resource with which to calculate intermediate milestones or checkpoints even before you feed your schedule to a computer. This will provide you with a quick check of the accuracy of your computer output. This facility will also provide you with the resource to make rapid checks of schedules provided by architects and contractors, schedules associated with changes, and updates of progress schedules. You might even become facile enough in early project meetings to convince those who contract with you to provide you with no-nonsense schedules.

As more and more work items and relationships accumulate with the development of this network analysis, it becomes quickly apparent that some work items can be performed over longer periods, or can be delayed, without affecting the overall duration of the project. The aggregate durations of the work items on these *paths* can be compared with the aggregate duration of the *critical path* so that the difference can be calculated. This difference in these aggregate durations is the *additional time* that can be expended on these paths of *noncritical* work items without delaying the project. It is called *float* or *slack*.

> It is important to note before going further that *float* is associated with the path on which these work items fit. It does not relate to individual work items; i.e., if time is lost on one of the early work items on a *float* path, the following items on that path will have less float remaining.

This consideration alone (there are many others) underscores the importance of regularly updating CPM schedules in order to represent both relationships and actual events accurately.

Now, it's time to have a little fun with CPM.

The Hargroves Build a Backyard Fence

The Hargrove family have just gotten a puppy. Right now, they don't have a fence around their backyard, but they don't want *Max* to get out into the street. So, they've decided to build a white picket fend around the back of the house. Mom (Camille) and their teen-age daughter Vickie will keep the men fed and might be *"go-fers,"* but most of the work will fall to Bill and son David.

Bill and Dave would like to get the fence done in one weekend. This is their first pass at projecting a schedule for the work: They use hours for their schedule of work. And they put in a gap for overnight, equivalent to a weekend on a construction project.

Trial No. 1

On their first pass, Day 1 takes up 8-3/4 hours. Day 2 takes up 6-1/4 hours before they have to quit, so the concrete they've used to set the posts will cure enough for them to set the rails.

That puts them over into Day 3, which takes another 8-1/2 hours. Note that they've had to take a break of two hours in the middle of the third day so the base coat of paint can dry.

How can the Hargroves get the project down to a two-day weekend without working 16-hour days?

Some suggestions:

- The Hargroves got *Max* on a Friday. Bill could gain 45 minutes on Saturday by sketching the fence on the dinner table Friday night.

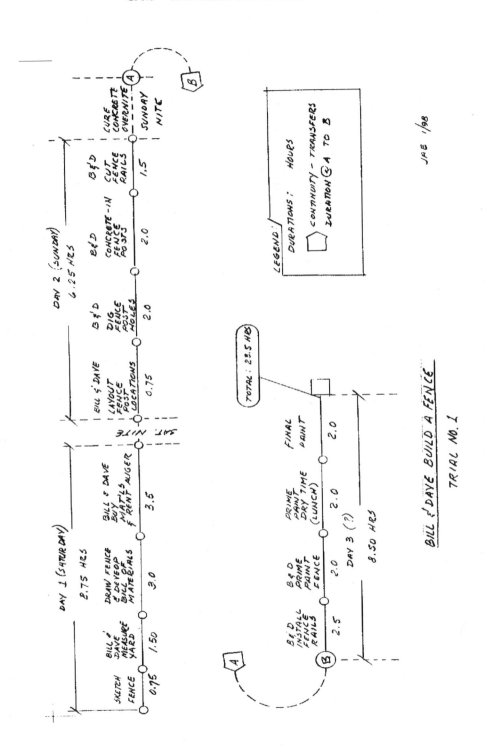

BILL & DAVE BUILD A FENCE

TRIAL NO. 1

DAY 1 (SATURDAY)
8.75 HRS

SKETCH FENCE — 0.75
BILL & DAVE MEASURE YARD — 1.50
DRAW FENCE & DEVELOP BILL OF MATERIALS — 3.0
BILL & DAVE BUY MAT'LS & RENT AUGER — 3.5

SAT. NITE

DAY 2 (SUNDAY)
6.25 HRS

BILL & DAVE LAYOUT FENCE POST LOCATIONS — 0.75
B&D DIG FENCE POST HOLES — 2.0
B&D CONCRETE-IN FENCE POSTS — 2.0
B&D CUT FENCE RAILS — 1.5
CURE CONCRETE OVERNITE — SUNDAY NITE

A → B

DAY 3 (?)
8.50 HRS

B&D INSTALL FENCE RAILS — 2.5
B&D PRIME PAINT FENCE — 2.0
PRIME PAINT DRY TIME (LUNCH) — 2.0
FINAL PAINT — 2.0

TOTAL: 23.5 HRS

LEGEND:
DURATIONS: HOURS

CONTINUITY — TRANSFERS
DURATION @ A TO B

JAE 1/98

405

Trial No. 2

- Looking ahead, it's evident that following the same sequence of work would still leave Bill and Dave waiting on the concrete to cure at the end of Day 2. So a new pass is in order. Time for the eraser.

- Let's have Dave get the post-hole digger (gas-powered) first thing Saturday morning, while Bill and Vickie measure the yard and locate the posts. Then, they can dig the holes as soon as Dave returns, and place the posts in the concrete Saturday afternoon.

- In Trial No. 2, Day 1 is 7-1/2 hours, but the posts are in the ground and curing overnight. Day 2 is 9 hours long by the time the rails and posts are prime painted, and they still need 2 hours to dry before the final coat. So the final paint for the rails have to wait for another day. *Or do they?*

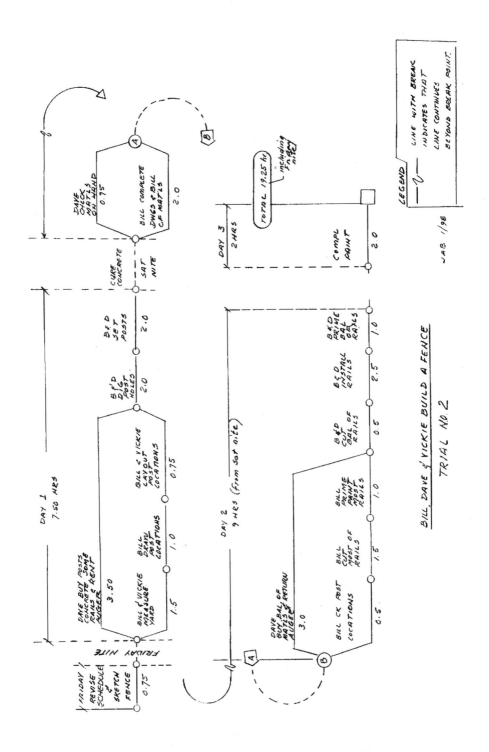

BILL, DAVE & VICKIE BUILD A FENCE

TRIAL NO. 2

A-2

Trial No. 3

- On Day 1, Vickie and Camille prime paint most of the rail material (before it's cut) while Dave and Bill place the posts in concrete. Camille and Vickie help mix concrete to expedite the placement of posts. Then, the concrete can cure, and the prime paint can dry Saturday night.

- On Day 2, Bill can get some help from Vickie checking post locations (for accurate cutting of rails) and together they install the first rails. Dave, meanwhile, gets the rest of the materials (including some more rail material, if necessary) when he returns the auger.

- Vickie prime paints the remaining rail material while Bill and Dave install the rest of the rails, and there's time enough for it to dry and to apply the final paint on Sunday afternoon.

The project that originally filled 23-1/2 hours came down to 16-3/4 hours, with rescheduling and some additional resources. Simple as this project is, it starts to reflect what a CPM graphic analysis and an active eraser can do on more complex projects.

Before you proceed to the short exercise below, you might want to develop your own network planning for a family vacation, or a more complicated project involving several people or teams.

A Valuable Resource:
How to Manually Calculate a CPM Schedule

On your building project, this is the kind of effort that will help you evaluate alternative approaches to planning your project, even before you feed your analysis to a computer. It will also assist in assessing the reasonableness of others' schedules.

It's important, before you light up a computer to calculate all the cumulative durations and dates in a complex schedule, to have identified the controlling overall durations (the "critical path" or paths) and to have calculated the cumulative durations to various milestones. These will provide checkpoints against which to check the accuracy of computer printouts.

Since you will want to check your own and others' schedules against a calendar, you can use the computer, first, to produce a table that correlates working days to calendar dates over the pro-

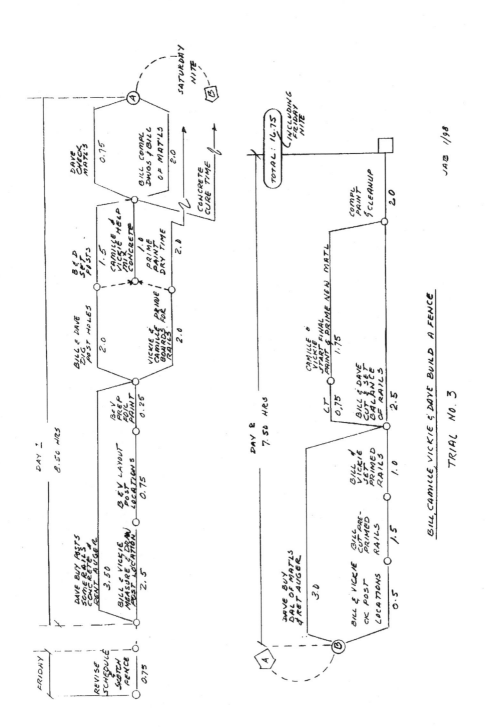

BILL, CAMILLE, VICKIE & DAVE BUILD A FENCE

TRIAL NO. 3

JAB 1/98

A-2

jected duration of your project. All competent scheduling software can produce a date-correlation table.

As an exercise, you might want to try your hand manually calculating the durations and floats of the diagram on the next page.

It's a more complicated network diagram—with 22 work activities, a lead time and various dummies—than you've seen so far. The first sheet shows only the durations, but no indication of where the critical paths run or the float of any activity or path.

As indicated by the Legend, you can show the Early Days above the line, and the Late Days below the line, with the float in brackets below the assigned duration. The solution is shown this way on the next page.

Why Is CPM Important?

The measurement and control of how *time* is planned and used on a building project are of sufficient importance to warrant this emphasis:

No resource is more important to the owner of a construction project than *time*.

The abuse of time during either design or construction can add up to cost overruns more quickly than almost any other problem, short of an outright disaster.

The misrepresentation, in support of a construction claim, of how time was used on your project is rarely understood or challenged by those on whom you may rely for objective judgments.

If you are to understand and respond to how time will be, or was, used on your project, it will be essential to have in your toolkit a facility with which you can challenge the kind of misinformation that passes as intelligence on others' projects.

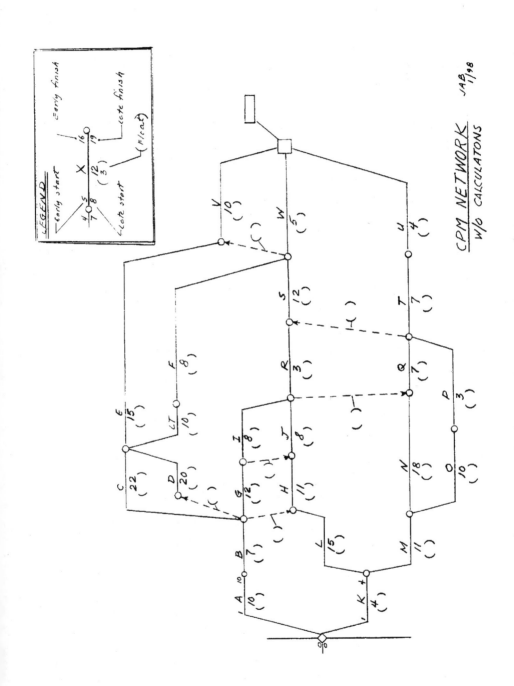

CPM NETWORK
w/o CALCULATONS

JAB 1/98

A-2

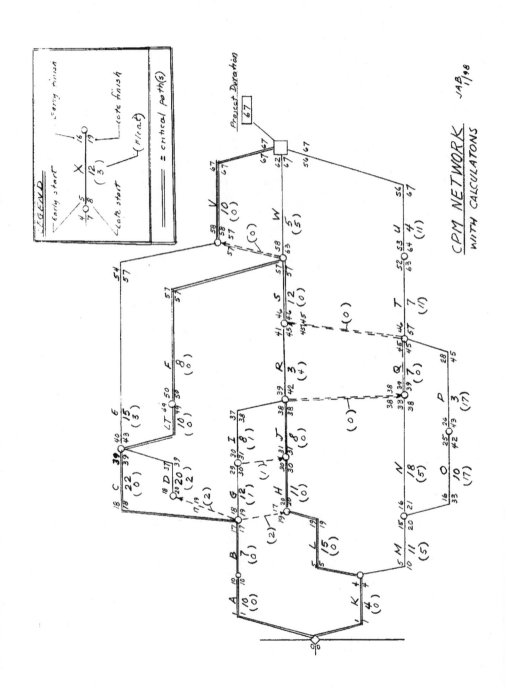

412

POST SCRIPTS TO THE USE OF
CPM IN CONSTRUCTION

Owner Supported and Maintained CPM System

If you have recognized the merits of reporting and measuring time in construction, then it's worth considering implementing the system JBA recommended to a federal agency in 1969 on this basis:

- The owner owns the project. Time is one of the project's most important resources. Therefore, it is well worth the investment for the owner to define the scheduling system to be used on the project (including a specified software in which it is to be processed), so the owner can process scheduling information in parallel with the requirements incorporated into the architect's and contractor's Contracts.

- This eliminates the difficulties associated with trying to understand how to read someone else's arcane system. Since the data developed in a network analysis can be processed equally on almost any system, this does not impose an undue requirement on either contracting party.

- This provides the owner the facility to analyze changes *before* presenting them to the contractor to request proposals, and after asserted delays and/or claims are presented. It facilitates negotiation of time-related change orders.

- If the owner has multiple projects, it standardizes the owner's scheduling system across the board.

- It provides an archiving system for accumulation of project data, and for training of owner staff.

- It circumvents the excuse *"our scheduler is sick,"* or *"we're pretty much on schedule so why don't we bypass updating the schedule?"*

Project-Delay Claims

If the owner has maintained a parallel scheduling system as the project advances, the piece of junk described on page 211 is easy to shoot down.

413

If it has been maintained in parallel with a computer database that can be used to access project documentation, and particularly if this continuing process is demonstrated by the owner's representative(s) in meetings, most contractors will recognize the futility of asserting delay claims without reasonable justification.

This parallel tracking of progress and performance eliminates both the gratuitous opinions of those experts, and the obstructions to discovery we discussed back on page 164.

This approach both encourages and facilitates negotiation of time-based disputes without recourse to the claims process (and consequently without attorneys and "other strangers").

Conversion (If You Must) of the Arrow Diagram to the Precedence System

One day, someone will try to convince you that you should be using the *precedence* system for scheduling.

As discussed previously, this approach started by showing *time in a box* and connected boxes with lines that showed their relationships. It was asserted—together with the idea that it would reduce the load on computers—that it was easier for most people in the field to understand.

In a word: baloney. It was easier only because most early precedence systems were so simplistic there were few work items to evaluate.

It was also "easier" because it was possible to run the lines cattywampus backwards and forwards all over the diagram to correct or adjust oversights and afterthoughts. As such, it eliminated the rigorous thinking that an arrow diagram *with all arrows pointing to the right* imposes on the analysis of the processes being scheduled.

Now, most precedence systems are drawn much like arrow diagrams—so, developed with the basic rule that a work activity is defined every time someone passes work off to someone else, they can be as rigorous as an arrow diagram.

But in the meanwhile the computer software that's been marketed has stuck with the same single-numbering system and awkward annotations we discussed before, in lieu of showing *dummies* and *lead times* to reflect meaningful relationships.

What has been lost is much of the original benefit of putting the relationships into Mr. Gantt's barchart: the relationships are hidden in computer notations, and only visible through the reading of difficult listings of "predecessor" and "successor" activities.

The very real value provided by the arrow diagram—the value of *graphically demonstrating* all work item relationships—has been lost. The consequences are the absurdities you've read about in the war story on page 252.

However, if you become convinced, you will find little difficulty having your arrow diagram translated into data that can be processed in a *precedence software scheduling system.*

Fragnet

How the industry loves buzzwords. This word usually describes a fragmentary part of a network analysis that is added to an existing schedule to test how it may affect the scheduling of the Work.

Apart from the use of the buzzword "fragnet," it is the practical use of this idea that was part of what I proposed a couple of pages back: that the owner maintain his own updated version of the project schedule independently of what the contractor reports.

Then, when changes occur, or when the owner intends to propose a change, he can first develop a "logic fragnet" of that change and insert it into the currently maintained CPM network.

He will then have a resource with which to take the initiative in negotiating disputes, or negotiating the reasonable insertion of changed work into the project.

JOE BOYD

The insights Joe Boyd has incorporated into *Build It Twice* reflect his more than 40 years in the construction industry, and his disposition to view problems as challenges and opportunities.

His insights reflect the resources he accumulated at Santa Clara University, where, in addition to graduating in business and accounting, he managed the university newspaper and taught journalism courses.

They reflect two years of graduate law until the army called him up during the Korean War to teach cryptography and for assignment to SHAPE Headquarters, Paris.

They reflect his return to the University of California at Berkeley, where he obtained a degree in civil engineering.

They reflect his experience in design offices, in the field with general contractors, as senior engineer and superintendent with the Bechtel Corporation and his position as Director of management services with Transamerica Corporation.

By the time he established his own firm in 1963, Joe had already become an expert in the use of CPM, in cost and claims analysis, and had assisted with the management and claims resolution of more than $150 million of construction.

J.A.Boyd & Associates (JBA) grew to a staff of 35 and extended its services throughout the U.S., providing consulting and management services on several billion dollars of design and construction, and for the analysis, negotiation and resolution of several hundred million dollars of construction disputes.

In addition to directing the services of his firm, Joe Boyd has testified as an expert, and provided services as arbitrator, referee and negotiator on over a hundred diverse construction disputes.

ACKNOWLEDGEMENTS
To All of You Who Made This Book Possible...

BIT would not exist without your many, constructive and diverse, contributions.

The largest number of you, however, must remain nameless, since it's your histories, your agonies and the absurdities you permitted that have inspired the stories—generic though they are—that put flesh on the bones of the ideas here.

Then there are those of you who for several years, or many, contributed to the experiences we shared at J.A.Boyd & Associates. While we were usually about 30 to 35 at any one time, capabilities, limitations, ambition, the rapine of competitors and natural attrition meant that over a hundred of you transitioned through JBA over 35 years. Many of you made substantial contributions to JBA and to our clients, but one of you, particularly, is at the core of *BIT*.

John Schagen, AIA, joined us in 1969 and retired in 1993. It was his frustration with the members of his profession, and his incredible ability to penetrate the ambiguities hidden in their construction documents that showed us the possibilities of implementing effective B/BAs.

Thank you, John, and those of you whose loyalty, standards, perspicacity and perseverance reduced to surmountable challenges the problems of dealing with naïve and reluctant clients, and truculent "other strangers."

Then there are those of you out in the industry and the professions who demonstrate that it is still possible to find and to work with people of competence, integrity and commitment. Your contributions, comments, critiques and edits have improved *BIT* beyond measure. While it still contains the shortcomings I've contributed, I need to thank, in particular, Jim Luce, Malcolm Misuraca, Mike Cannizzaro, Greg Thomas, Dave Baldwin, Patricia Schwafel, and Jim Kwapil. To those of you who have taken the content of this book to heart, my hope is that you find *guides* equal to these.

Then, there is one virtue for which I need to thank everyone associated with the production of *BIT*. Your patience. It has gone through many drafts, and all of you have borne with exceptional equanimity my constant re-writings.

Beth Rodda, since she has been with it from its early drafts, not only contributed its layout and book design, but has suffered through the typesetting and reworking of endless revisions. Her stick-to-itiveness has carried it even through its final printing. Thanks to Michele Le Blanc for her excellent illustrations. And continuing thanks and kudos to Five Star Publishing Support—in particular Sal Caputo, Linlie Hermann and Linda Radke—for their assistance, respectively, with editing, cover design, PR and marketing.

Gloria Casas, executive assistant, girl-friday, editor and general factotum for many years with JBA, and now with SRP, continues to demonstrate her fortitude for a boss and an author constantly trying new things, whatever their merit.

At the heart and soul of everything here, not only *BIT*, but for the past 34 years, is the vice president of JBA, office manager for its entire crew, constructive critic, constant, supportive companion, contributor of insights, workshop assistant, co-pilot, co-traveler, conversationalist, best friend and wife, DeAnna.

BIT is the culmination of an effort that would not even have been considered had DeAnna not been watching attentively for things that always seem to arrive out of left field, and had she not been deflecting almost from the day we met, the nuisances that plague all of our daily lives.

Quotes on pages 302 and 303: 21st Century Leadership, by Senn-Delaney Leadership Consulting Group, Inc. pp. 96, 123, 147-48 and 211. © Copyright (1993) by Leadership Press, Inc. Reprinted with permission. Further reproduction prohibited. All rights reserved.

Quote on page 82: S, M, L, XL p xix. © The Monacelli Press, Inc. Used with permission. Further reproduction prohibited.

Quotes on page 302: Partial quotations reprinted with permission of Simon & Schuster from THE BOOK OF VIRTUES by William J. Bennett. Copyright ©1993 by William J. Bennett.

ORDERING INFORMATION

Telephone: (707) 257-5292

FAX: order form to (707) 226-1438 with credit card information

Mail: Split Rock Publishing Company • P.O. Box 5517• Napa, CA 94581

— —

Please send me _____ copies of *Build It Twice . . . If You Want A Successful Building Project* at $30 per copy plus $4 for shipping first book ($1 each additional book). Shipping via Priority Mail or UPS Ground Tracking. UPS requires a street address, they cannot ship to a P.O. Box. Other shipping options available upon request.

Check One: ☐ Cashier Check or Money Order[1]   ☐ MC[1]   ☐ Visa[1]
☐ Personal Check[2]

[1] Allow 2-3 week delivery of orders made by cashier check, money order, or credit card.
[2] Allow 4-6 week delivery of orders made with personal checks.

Card No.   _____/_____/_____ Exp. Date:____/____

Signature: _____
(not valid without signature)

Make check/m.o. payable to: **Split Rock Publishing Company**
California residents add appropriate local sales tax.

SHIP BOOK(S) TO:

Name/Company _____

Address _____

City _____ State _____ Zip _____

Day Telephone: (_____) _____-_____

E-mail _____

Split Rock Publishing Company • P.O. Box 5517• Napa, CA 94581 • FAX (707) 226-1438
e-mail: bit@splitrockpub.com